Woolly

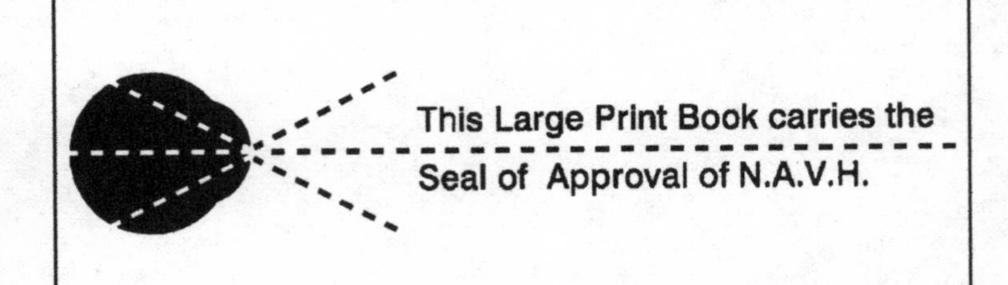
This Large Print Book carries the
Seal of Approval of N.A.V.H.

WOOLLY

THE TRUE STORY OF THE QUEST TO REVIVE ONE OF HISTORY'S MOST ICONIC EXTINCT CREATURES

BEN MEZRICH

Epilogue by Dr. George Church
Afterword by Stewart Brand

THORNDIKE PRESS
A part of Gale, Cengage Learning

Farmington Hills, Mich • San Francisco • New York • Waterville, Maine
Meriden, Conn • Mason, Ohio • Chicago

Thorndike Press, a part of Gale, Cengage Learning.

Thorndike Press® Large Print Popular and Narrative Nonfiction.
The text of this Large Print edition is unabridged.
Other aspects of the book may vary from the original edition.
Set in 16 pt. Plantin.

LIBRARY OF CONGRESS CATALOGING-IN-PUBLICATION DATA

Names: Mezrich, Ben, 1969- author.
Title: Woolly : the true story of the quest to revive one of history's most iconic extinct creatures / by Ben Mezrich ; epilogue by Dr. George Church ; afterword by Stewart Brand.
Description: Large print edition. | Waterville, Maine : Thorndike Press, A part of Gale, a Cengage Company, [2017] | Series: Thorndike Press large print popular and narrative nonfiction |Series: Include bibliographical references
Identifiers: LCCN 2017026573 | ISBN 9781432841072 (hardcover) | ISBN 1432841076 (hardcover)
Subjects: LCSH: Woolly mammoth. | Extinct animals—Cloning. | Extinct animals—Genetics. | Large type books.
Classification: LCC QE882.P8 M49 2017b | DDC 591.68—dc23
LC record available at https://lccn.loc.gov/2017026573

Published in 2017 by arrangement with Atria Books, an imprint of Simon & Schuster, Inc.

Printed in Mexico
1 2 3 4 5 6 7 21 20 19 18 17

To Asher and Arya,
who will grow up in a world filled with
Woolly Mammoths

AUTHOR'S NOTE

Woolly is a dramatic narrative account based on numerous interviews, multiple first-person sources, and hundreds of pages of articles. In some instances, settings have been changed, and certain descriptions and characters have been altered to protect privacy. I employ the technique of re-created dialogue, based on the recollection of participants who were there, various diaries and files, and newspaper accounts, doing my best to communicate the substance of these conversations, especially in scenes taking place years ago.

■ ■ ■ ■

PART ONE

■ ■ ■ ■

Very often, as I wander through life, I'll get that old feeling that I've come back from the future, and I'm living in the past. And it's a really horrible feeling.

— GEORGE M. CHURCH

The rewards for biotechnology are tremendous — to solve disease, eliminate poverty, age gracefully. It sounds so much cooler than Facebook.

— GEORGE M. CHURCH

Chapter One

Three thousand years ago

WRANGEL ISLAND.

An eighty-mile-wide swath of volcanic rock, gravel, and permafrost jutting out from the Arctic Ocean, ninety miles off the coast of Siberia, windswept and forbidding. Gray on gray on gray, a forgotten stretch of beach covered in a thick mist, the air heavy with the spray from waves crashing through the loose ice beyond the breakers.

A little after 5:00 a.m., the calf opens his eyes.

Even though his mother is only a few yards away, splayed out against a makeshift nest of dry reeds that she has gathered from the underbrush deeper toward the center of the island, the calf feels strangely alone. The rest of his herd — twenty-five strong, arranged along a matriarchal line that stretches back through three generations — has already begun a short pilgrimage down the coast. Being separated from the bulk of the herd feels unnatural. A mild illness has briefly kept his mother from engaging in the routines of survival on the island, so she's stayed behind as the herd has set out to search for more sources of drinkable water and grazing. At less than a year old, the calf has remained with his mother, his familial bonds and youthful needs overcoming his developing social instincts. But neither nature nor nurture makes sitting around waiting for her to get back to her full energy any easier.

The calf pushes himself to his feet, the thick muscles in his enormous legs trembling with the effort. His size already makes rising from the ground a bit of an ordeal. He weighed over two hundred pounds at birth and even then stood over three feet tall. Now, though he is far from fully grown, he weighs well over a thousand pounds.

He shakes his head back and forth, shedding reeds and bits of snow and ice that gathered over him as he slept. His mother is still lying on her left side against the ground in front of him, her huge body rising and falling as each breath sends clouds of condensation through the frozen air. As big as the calf is, his mother is a veritable mountain, five, maybe six tons, and more than twice his height. It is no wonder that his kind often naps standing up. When they do sleep flat against the ground, it is usually for periods of no more than four to five hours.

The calf watches his mother for a few minutes, then kicks the last bits of ice off his legs and starts forward down a gentle, gravelly slope that leads toward the beach.

Each heavy step sends tremors up and down his body, as his huge feet compress and churn the frozen ground. The wind howls around him, pushing his tiny, round ears flat against the sides of his head, but he continues forward, his eyes searching the turned permafrost beneath him for bits of grass, moss, roots. As he nears the bottom of the slope, he starts to feel the spray from the waves that crash against the large volcanic rocks making up much of the coast; the water feels good against his body, the glistening, bluish drops beading against the long strands of thick red hair that cover most of his hide.

Despite the wind, the icy water, the cold, the calf is not uncomfortable. Though it is a harsh environment, he and his herd are uniquely suited for it; in fact, for thousands of years, Wrangel Island has *enabled* the huge animals to survive and thrive.

Even now, his is one of perhaps two dozen herds on the island. At one time, the super-herd numbered close to a thousand individuals, though in recent years it has dwindled to half that.

Though the interrelated herds have always lived in proximity to various predators, it wasn't claws and teeth that cut down the calf's cousins, aunts, and uncles in recent years. The thinning of his kind was part of a natural process of adaptation. The world around him has changed, and his species has adjusted; smaller in number, leaner, but functioning. In this forgotten corner of the world, they have learned, survived.

In fact, though the calf couldn't possibly know, the isolated, icy nature of his island home is the only reason the herd still exists at all. A twist of fate, an accident of geography, a turn of weather: Six thousand years earlier, as the world had first begun to warm, the water surrounding Wrangel had risen — fifty feet or more — and cut off the island from the mainland. The calf's ancestors, who had crossed over one herd at a time along an ice bridge during the colder months of the year, had found themselves trapped. Lost in time.

Saved.

While the calf's super-herd adapted to its

isolation on Wrangel, the rest of their species had died off around the world, over four thousand years ago. The five hundred or so individuals left on the island are all that remain. Split into familial units, close-knit, living in a symbiotic relationship with the island itself, they have lived four thousand years beyond their kind's extinction.

The calf finally reaches the bottom of the slope and is now less than a dozen yards from the water itself. The spray is even more palpable now, the frozen droplets pelting his face and hide like hail. It is time to head back up the slope toward his mother. Perhaps she is awake now and well enough to finally rejoin the herd, farther down the coast. He starts to shift his heavy body in the opposite direction, when something out on the water catches his attention.

Cutting through the waves, slicing past the jagged chunks of ice and over the breaking foam — something the cub has never seen before. He stands frozen in place, staring at the long, cylindrical object, which his mind had no capacity to understand.

Like a hollowed-out tree trunk, the object lies horizontal, moving forward on the ocean's surface toward the beach — right in his direction.

The calf takes a step back, then freezes

again. Above the edges of the long object, he can now make out creatures, five or six of them, huddled together against the water's spray. They are small and pale, covered in odd hides that aren't hairy. And they are pointing at him.

He watches as one of the creatures rises and lifts a thin wooden shaft tipped in razor-sharp bone high into the air. It is twice as long as the creature itself.

The calf stares, too stunned to move. He does not know what these strange creatures are, or why they are heading to his beach. He cannot know that they have come to finish what the millennia of a warming world have not.

The calf cannot know that he, his mother, his herd, are the last of his kind.

After him, there will be no more.

Chapter Two

Four years from today . . .

CHERSKY, SAKHA REPUBLIC, SIBERIA.

Justin Quinn was breathing hard as he trudged behind his long-legged Russian guide, trying not to trip over the patches of brush that randomly sprang from the fracturelike cracks speckling the frozen tundra beneath his snow boots. The boots weren't his; he'd borrowed them from the locker room at the Northeast Science Center in Chersky four miles back, and though they were much warmer than the pair of hiking boots he'd brought with him from Boston, they were at least two sizes too big, the rough material reaching almost to his knees. Nor did it help that the parka he was wearing — also borrowed — was also oversized. The heavy synthetic fur of the hood kept getting into his mouth, and the sleeves went to his fingertips, making him look like a slovenly child, rather than a twenty-nine-year-old postdoc from the most prestigious university in the world.

"It's not much farther," the Russian called back over his shoulder, his English tinged with such a heavy accent that Quinn had to concentrate to understand what the man was saying. "You're doing much better than the last graduate student they sent me. He barely made it past the first mile marker."

Quinn wasn't surprised. Even with the borrowed gear, every breath he took of the frozen air sent knifelike shards of pain through his lungs, and the skin of his face that was still exposed beneath that damn cocoon of fur was already numb. Quinn wasn't a waif by any means; he'd been an athlete in high school and the various colleges he'd attended, before his academic interests made daily three-hour practices an untenable use of his time, but he wasn't anywhere near fighting shape. Years sitting in various labs in Boston playing with test tubes, Petri dishes, and centrifuges would do that to you. And even if he'd been in peak condition, the terrain they had crossed since leaving the institute would have had him breathing hard.

To call this stretch of geography "forbidding" would have been a gross understatement. Barely ninety miles from the Arctic

Ocean, technically well within the geographic zone known as the Arctic Circle, the steppes surrounding the Northeast Science Center consisted mostly of windswept hills made of permafrost — ground that never thawed, up to thirteen feet deep in some places, covered by a thin layer of rocky soil — out of which sprouted various weeds, mosses, and lichen. Even though it was mid-April, the temperature hovered at a balmy −4 degrees Fahrenheit. And as frigid as it seemed, Quinn knew he was catching the region on a good day; the Republic of Sakha, the area of Siberia where the Science Station was located, was generally known for being the coldest section of the northern hemisphere. During the winter months, the temperature could drop as low as −76 degrees Fahrenheit.

Stumbling forward over the ice-hardened ground, the wind biting at his cheeks, Quinn had a hard time believing anything could survive this far north. Then again, his Russian guide didn't seem to mind the cold; in fact, he seemed to be moving faster the farther away they were from the relative comforts of the Science Station. Before he'd first packed up and headed into the Russian wilderness, Quinn had been told that his guide had been living exclusively at the station for more than two decades. In fact, the Russian and his family had secluded themselves in this stretch of harsh ecology to such a degree, some might have questioned their sanity. But that was often the case with the brilliant and the obsessed.

"You see, we are just about there."

The Russian pointed a dozen yards ahead, toward the edge of what appeared to be a large, fenced-in enclosure. About ten feet high and made of a sturdy, chain-linked metal, the fence ran half the length of a football field in either direction. As Quinn continued to trudge closer, he saw at least a dozen shapes on the other side of the links. Animals, standing in groups of three and four, seemingly immune to the cold and the wind.

Quinn assumed they were horses. At least they were horse shaped. But they were covered in thick brown fur and were muscled and stocky. They were not quite as short as a Shetland pony, but smaller than the horses Quinn had seen in America.

"What are they?" he asked, as he trailed the Russian, who was walking parallel to the fence. "I don't think I've ever seen anything like them before."

"Yakutian horses. We have nearly two dozen now. The last shipment came in three months ago and they've acclimated quite well. This is temperate for them, actually. In fact, they've even started to breed."

"Wow," Quinn managed to say. He nearly tripped over a jagged piece of ice sticking up through the middle of a bed of frosted moss, but caught himself as they continued toward the far corner of the enclosure.

"Yes, wow." A very old breed, Yakutian horses evolved to live in the cold of the Yakut region, where they can locate vegetation that is under deep snow to graze on.

The Russian pointed to his right, in the direction of another fenced enclosure, a few dozen yards away.

"Over there, we have a group of North American bison that were shipped over last year. Beyond that, there's a group of reindeer from Finland. Most of them were in pretty poor shape, but much cheaper than the bison. Hopefully the more persuasive our data become, the easier it will be to gather more specimens."

They'd reached the far corner of the enclosure, moving close enough that Quinn could have taken off a glove, reached out, and touched the metal links, if he hadn't been worried about losing a finger to frostbite.

"From what I've read, the data's beyond

persuasive."

"Here you don't need to read, you can simply look and see for yourself."

The Russian pointed at the ground where the furry, stocky horses were grazing. Then toward the ground beneath Quinn's giant boots. Quinn didn't have to be a scientist to notice the clear difference: Outside the enclosure, little life beyond sparse moss and lichen was growing, but within the enclosure, where the Yakutian horses were huddled, the ground was covered in patches of thick, green grass.

"When we get enough animals and take away the fences, the data will become even more extreme."

Quinn would have whistled, but his lips were nearly frozen. He knew, from his pretrip reading, that it wasn't just the handful of animal enclosures at the core of the Russian's rapidly expanding data set. Yakutian horses, bison, reindeer — the resettled animals were just one facet of the scientist's ambitious experiment. The Russian's team — beginning with his father, the true genius behind the experiment — had been working the acres of land around Chersky for decades now, with tools ranging from repurposed bulldozers with heavy, spiked treads, wheeled pile drivers that could pound the ground with precise, incredible pressure at the push of a button, to a surplus issue, World War II–era tank they had purchased from the Russian government in Yakutsk and driven several hundred miles to Chersky.

Between the animals, the construction machinery, and the tank, they'd managed to accomplish something that many scientists might have thought impossible: Beneath the layer of grass within the various enclosures — the controlled test environments — they'd lowered the permafrost ground temperature by an average of fifteen degrees.

Quinn wasn't a climatologist — he was actually a student of biology, with a background in genetic engineering — but even he knew these numbers were staggering; more than that, they were important.

Important enough to inspire him to take a trip halfway around the world.

"When we take away the fences," the Russian continued, without slowing his pace, "it will be like turning back time ten thousand years. Specimens will become herds. A repopulation."

"And what about predators?" Quinn asked, as they passed the edge of the enclosure and started slowly to ascend a hill of permafrost. "I assume they were once part of the history of this place?"

"An important part, yes. Arctic wolves, polar bears. Before that, saber-toothed tigers."

"Saber-toothed tigers?"

"The predators and the herbivores lived side by side. In fact, the predators helped the herbivores to thrive. They were territorial, and protected their herds from encroaching competitors. They also picked off the sick and weak. It was an important balance. Until we came along."

Quinn knew where the story went from there. As the last ice age ended, human populations moved northward. Hunting parties decimated the native populations of herbivores and predators. Though the science was still a bit controversial, the Russian experiments — the resettled animals, the tank turning over the soil, the pile drivers — were meant to provide proof that it wasn't just the changing environment that had caused the mass extinctions. *In large part, it was the other way around.*

"At first, we were like every other predator in this place. We hunted what we needed. But we never get full. We never stop hunting. We aren't just another predator — we are the Apex Predator."

The Russian looked back at him, as they

reached the top of the slow rise. Quinn could see past the Russian's shoulders; ahead of them rose another fence, at least twice as high as the Yakutian horse enclosure. It was also made of metal, but this fence was much thicker, and topped in curls of barbed wire. It took Quinn a full minute of breathing hard to realize — this fence wasn't built to keep predators out. It was built to keep something in.

The Russian strolled forward, toward a chain-link door built into a section of the fence directly in front of them. There was an electronic keypad halfway up the door, covered in Cyrillic letters. The Russian punched six of the keys in sequence, and they heard a metallic whir, electronic tumblers falling into place. Then the door swung inward. The Russian gestured for Quinn to step through.

"Is this safe?" Quinn asked.

"You don't sound much like an Apex Predator."

Quinn swallowed, hard.

"Actually, I'm a vegan."

The Russian laughed.

"It's perfectly safe. I promise, no saber-toothed tigers. Nothing but herbivores."

The Russian waited for Quinn to go through the open doorway first, then followed behind. Quinn found himself standing at the top of a long incline, looking down at an expanse of steppe similar to the four miles he'd just hiked — but, beginning a dozen yards ahead of him, much of the moss and lichen had been replaced by thick, green grass. Something had been turning this topsoil with enough regularity to make a real, significant difference. From the size of the area within the fencing, Quinn knew it had to be something much bigger than a World War II tank.

He was about to take another step forward when he saw something in the distance, moving toward them. The shape was large, lumbering, and vaguely familiar. But as it continued toward them, step by massive step, it seemed to keep getting bigger.

Too big, Quinn suddenly thought to himself.

Much too big.

"Christ."

"Not this time."

"This isn't possible . . ."

But Quinn knew better than anyone, that wasn't true.

Although more than five years had passed since he'd left the lab — and the team associated with this place at the northern edge of the world — and had rejoined only a week before heading to Russia, he knew that his colleagues had only a few rules that couldn't be broken: Highest on that list was that the word "impossible" was officially banned.

During his time at the lab, Quinn had seen enough incredible things to know why. On his return, his colleagues had purposely kept him in the dark as to their progress, wanting him to see it for himself, firsthand. He knew that a thing like this — the creature he was seeing, something that shouldn't have existed, that *hadn't* existed for more than three thousand years — wasn't simply possible.

It was inevitable.

Chapter Three

Today

77 AVENUE LOUIS PASTEUR, BOSTON.

Ten minutes past two in the morning, and the warrenlike lab tucked into the second floor of the glass and steel New Research Building at Harvard Medical School was as alive as the middle of the day. Teams of young postdocs, grad students, and harried fourth-year med students huddled over high-tech workstations, engaged in what appeared to be a highly choreographed dance involving pipettes, Petri dishes, and DNA-sequencing arrays. Gloved hands moved in and out of sterilization chambers and secure specimen freezers, and masked and youthful faces hovered over test tubes, twirling like small tornadoes within chrome-plated centrifuges.

In the middle of it all, Dr. George Church strolled through the beautiful chaos, a grin painted above his billowing white beard.

Science was supposed to be staid, boring, a slow drip of sap running down a tired maple tree. But even on a bad day, Church's lab was anything but dull, and tonight, the place was running at a hundred thousand RPM. All Church could do was stand back and watch, as his ever-growing litter of young charges raced toward yet another breakthrough of epic proportions.

One of the running jokes that often moved through the New Research Building was that nobody knew exactly how many people Church's lab actually employed. Over the years, Church had determinedly gathered an eclectic group of the smartest young scientists from all over the world, but even beyond the kids he'd recruited, his lab had what Church liked to call an "open door policy." More than once, brilliant thinkers had literally walked in off the street. If they'd been able to make an impression, Church had invited them to stay.

Whatever the true number was — Church himself put it at ninety-one — his lab was now home to many of the brightest young minds in genetics, biology, and molecular engineering. Furthermore, Church had given them free rein: access to nearly unlimited resources, liberty to chase their ideas wherever they might lead, and most important, the keys to a startling new technology that made reading and editing the DNA of any living creature nearly as simple as cutting paper with a pair of scissors.

His unique lab was on the verge of another astonishing breakthrough, yet it was a natural extension of the work he had been engaged in for most of his adult life. At sixty-three, he was considered one of the most brilliant forward thinkers around, having been involved in the inception of important scientific endeavors from the Human Genome Project to the current battle to eliminate malaria and to reverse aging by using genetic implants. Physically, he was also impressive: tall, imposing, with his long white beard, and a thick blast of hair rising up from his head like a snow-ridden halo. A towering figure, not just in the rarified world of scientists, Church was also one of the few lab rats to cross over into popular culture. From a recent visit to Stephen Colbert's television show, in which Church wowed the audience by producing a slip of paper that contained seventy billion copies of his most recent book, which had been converted into chemical code and implanted into a fragment of DNA no bigger than a period, to his recent coverage in the *New York Times* for organizing a meeting of top biologists who were planning technologies to synthesize a human genome and other large genomes. (The article noted that the meeting was "private," which was then reported by other outlets as "secret," leading many scientists to jokingly start calling their own meetings "secret meetings.") Church was fast becoming the face of the genetic revolution, an area of science that seemed to promise extraordinary advances from designer babies to immortality.

And now, as Church strolled through his labyrinthine laboratory on the second floor of the New Research Building, well past two in the morning — which was even more impressive because Church was well known to rise every morning at five — one of those breakthroughs was only moments away. Approaching a group of young charges huddled around one of the workstations, he could tell that something incredible was about to happen.

Church leaned over their shoulders to look at what they were doing. On the table in front of them was a small plastic dish containing a single drop of hemoglobin, suspended in an "organoid" in a sterile saline base. The miniature, three-dimensional conglomeration of cells, grown from a small piece of tissue, had been created specifically to mimic a small interior organ. Through the microscope, Church could see the tiny architecture of a working circulatory system in its micro-anatomy.

A young Chinese woman placed the dish onto a thin metal tray, then slid the tray into a flash freezer. Within seconds, the temperature in the freezer reached a life-killing freezing point, representing an out-door air temperature of sixty degrees below zero.

A minute later, the tray was retrieved, and the dish with the hemoglobin was placed under the lenses of a high-powered micro-scope. One at a time, the postdocs looked at the sample, a tense silence spreading from their corner to the rest of the lab. Then the young scientists stood aside so Church could take his turn.

Peering through the microscope, he could clearly see: The hemoglobin cells in the or-ganoid were still active, still able to release oxygen. Still alive.

Sixty degrees below zero, deep winter temperature in the Siberian tundra.

At such a below-freezing temperature, most animals' blood would have long ceased functioning optimally for oxygen release.

"It worked," the young woman said, her tone characteristically terse. She was in her midtwenties, and English was not her first language. In fact, she had learned English in Church's lab. As unique a setting for linguistic education as it might seem, given her history, it made perfect sense. More than half of her life had been spent in labs and lablike classrooms of various shapes and sizes, on both sides of the world. Science practiced to such an extreme might appear like magic to the uninitiated, but to her it felt routine. Still, despite her stoic tone, she knew as well as Church that what they had accomplished was extraordinary.

The cells in the dish represented a sea change in the process of science itself.

Science was no longer confined to studying, understanding, and explaining the natural world. It was no longer limited to reading the secrets and mysteries hidden within nature. Science was now capable of *writing* those secrets, down at the cellular level. Biology and genetics had gone from passive observation to active creation.

Whether the young woman realized it or not, it was a shift George Church had been working toward his entire life.

Chapter Four

Early summer 1959

DAVIS ISLAND, TAMPA, FLORIDA.

A few minutes past noon, the mercury was well over ninety-six degrees, the air so thick and humid it seemed to flow from the sun in shimmering waves. George Jordan — two months beyond his fifth birthday, one stepfather removed from his birth name, George Stewart McDonald, and still another stepfather away from the last name he would finally settle upon as an adult, Church — stood knee-deep in mud, hands jammed into the pockets of a frayed pair of overalls.

"You're gonna need to step back a little farther, George. I don't want to have to explain to your mom why you're missing a piece of your head."

George shifted his feet backward through the murk, his boots making a sucking sound as they splashed brackish water, painting cauliflower streaks up the denim covering his legs. The older boy, Charlie, watched him from the small clearing a half dozen yards away, already bent forward over a quartet of plastic cowboys, a shiny Bic lighter gripped between the fingers of his right hand. Charlie's clothes were even more tattered than George's, and his shirt was two sizes too small; his jeans had holes in them big enough for a rattlesnake to slither through — a real possibility, considering where the two of them had spent most of their time since Charlie had moved in with George and his mother and stepfather, not three months earlier. The shirt — along with most of his clothes, shoes, toiletries, and whatever else could be considered personal property — were hand-me-ups from George, part of the shared, bunk-bed-style life the two kids lived, and a constant topic of humiliation, which George certainly understood. At nine, Charlie was a veritable adult in the rural culture of the swampy backwoods surrounding Tampa, and compared to George, he seemed positively worldly. In three months, he'd already taught George how to shoplift cowboys from the locked bins at the local grocery store, how to steal cigarettes from George's stepfather's medicine cabinet, and even how to hot-wire their neighbor's pickup truck, though they hadn't yet dared to take the thing for a joyride.

Charlie also knew a lot about fireworks.

"This is an M-80. A Class C explosive, illegal in every state. It's got the power of an eighth of a stick of dynamite, so when it goes off, you might want to cover your ears."

George felt a familiar spike of anticipation move through him as he watched Charlie flick the lighter open, the little flame arching up through the hazy air of the swamp. The plastic cowboys strapped around the red, cylindrical firecracker seemed resigned to their fate, subjects in what George and Charlie liked to call one of their many "experiments for the greater good of humanity." Two years earlier, the Russians had launched Sputnik — the world's first satellite, an event that still dominated newspapers — and it was the least George and his older housemate could do to further the cause, by studying the effects of sudden combustion on plastic action figures.

Even at five years old, George was on his way to becoming a scientist. Although nearby Tampa was the second-largest city in Florida, George had not grown up in an urban or even suburban environment. He was a child of the canals, mudflats, bays, and swamps of the surrounding islands. Independent from almost the moment he could walk, he'd cut his teeth in the wild forests and high reeds, among the strange animals and insects that turned every inch of mud seeping around his boots into ecosystems of their own.

Most days, before Charlie had entered his life, George could be found down on his hands and knees in that mud, digging up bugs, crustaceans, even scorpions and snakes. Just six months earlier, he'd met his first rattler. More thrilled than scared, he'd sat in awe as the thing coiled in front of him, its tail a blur of sound and motion. He'd wanted nothing more than to understand how and why it could make such a sound, why it lived in this swamp, how it interacted with the world around them. A short time later, he'd found an insect graveyard in a clearing behind a mudflat, full of discarded exoskeletons hanging from vines and the limbs of low trees. At home that night, he'd used a set of encyclopedias to teach himself about metamorphosis, to try to understand what had happened to the insects, why they had left part of themselves behind.

Even before he knew what the word meant, George had fallen in love with science. But he was facing an uphill battle; he didn't have any scientists in his life. While attending Miami Law School, his mother, Virginia, had met, married, and then divorced his father, Stewart McDonald, a pilot, race car driver, and barefoot water skier. Born on MacDill Air Force Base, George had been reared by a strong, independent woman who was essentially math and science phobic. Even so, she couldn't deny his obvious interest in and aptitude for numbers and the scientific process. Every night, George came home babbling about another adventure, about his hours spent digging in the mud. Then he'd head straight for whatever books he could find, matching pictures to whatever he'd seen that day. Although he was mildly dyslexic — he had trouble seeing letters in their proper places in words — he could quickly teach himself from the pictures. Once, he'd found a large insect submerged in a pond, which he described as a "submarine with legs." He'd sealed it in a jar, but the next day it had disappeared. It wasn't until he'd opened the jar that he'd noticed the enormous dragonfly hiding under the lid. Even though he couldn't read the words in any of the books his mother had gotten him from the library, the pictures taught him what had happened: The submarine with legs had been the larva of the dragonfly. The thrill of that discovery had been so intense that George became convinced he had found what he wanted to do for the rest of his life.

But for the time being, as George's mother worked to establish herself in the legal profession, her son's classroom remained the swamps. When George was three, his mother had remarried, giving him a new last name and a half sister; she'd also begun to make it a habit of taking some of her legal work home with her, which culminated in her delivering to George a new playmate, Charlie — a nine-year-old juvenile delinquent whom a judge had been unable to place in a foster home. Charlie had immediately taken George under his wing, and the swamp-classroom became a place for progressively more adventurous "experiments."

"Ready?" Charlie shot George an evil grin, and the younger boy took another step back, putting his palms over his ears. Then he gave a little nod.

"Blastoff!" yelled Charlie as he touched the flame to the firecracker's wick, then charged back through the swamp to where George was standing. There was a brief pause — and then a flash of light, bright enough to bring tears to George's eyes. A loud crack echoed through the air, and pieces of plastic rained down into the swamp. Something hot touched George's shoulder, and he slapped at it with his hand, sending a tiny, smoking cowboy hat splashing into a puddle at his feet.

"Total annihilation!" Charlie shouted, clapping his hands together. But George was too far gone to respond. In his mind, the mini cowboys still raining down into the swamp were miniature Sputniks caught in the complex math of forced acceleration and gravity. He could almost see the numbers dancing in the air. For him, the real fun came after the smoke had cleared. For George, the best part of their experiments was trying to understand the how and why of them.

Although looking at the licks of flame, the melting plastic, and the small crater the M-80 had left behind, George had to admit — sometimes it was fun just to blow things up.

Chapter Five

December 23, 2006

SAKHA REPUBLIC, NORTHERN SIBERIA.

A twisting stretch of highway, somewhere between Irkutsk and Chersky.

Nikita Zimov hunched over the steering wheel of his borrowed two-door pickup truck, the muscles in his forearms tight as metal ropes as he fought to stay near the center of the serpentine swath of packed mud. The road — more of a path, because nothing out here truly qualified as a road — cut between steep inclines of craggy rock and jutting promenades of heavily tangled forest. The beams of orange from the truck's waning headlights were like sickly pale fingers trying desperately to reach out through the inky blackness — the sort of complete dark found only in a place that did not see sunlight for months on end.

"This is crazy, isn't it? I mean even to you? Or does this feel normal?"

Nikita did not pull his eyes from the front windshield to look at the young woman sitting in the seat next to him, instead keeping his focus on the snowflakes that flew, horizontally, across the weak beams from the headlights. Her face was pressed against the glass of her side window, and she was desperately trying to make out anything beyond the shadows of trees and cliffs.

"The fact that it's normal doesn't mean it's not also crazy," he answered, trying to sound more confident than he was.

Anastasiya was barely twenty years old, the same as Nikita. A classmate of his at the University of Novosibirsk, she was also much more than a classmate. Nikita would never have invited a simple classmate to this place, above the ring of the world.

"You get used to the darkness, and the snow, and the cold. It's the polar bears that keep you on your toes."

Anastasiya glanced at him and he laughed, and then she was laughing, too. It was good, the laughter; it masked the fear ricocheting through his veins. His anxiety didn't have to do only with the terrain and the conditions — although to be sure, it was kind of like piloting a submarine through a swamp. His fear had more to do with where they were going and why they were going there.

He was on his way home.

That thought alone would have been enough to twist his nerves into a spiral, but there was also the fact that he wasn't coming home alone.

Before he'd left Chersky and the Northeast Science Center five years earlier, Nikita could have counted the number of girls he had ever met on one hand. After escaping Chersky to attend one of the country's best science-focused high schools in Novosibirsk — then continuing to the university — he certainly had not expected to find someone whom he'd felt strongly enough about to chance bringing her to this place. Over the entire journey, he had half expected her to turn around and run. A girl who had never seen snow in her life, she had landed at that little airport in Irkutsk in the middle of the winter, then headed down the road with him into all this blackness.

And then of course there was the cold; they had stepped off that plane into a stiff wind that brought the temperature well below −40 Celsius. And yet, as he had retrieved the truck that his family had left for him for the two-hour trip to Chersky, she had simply smiled and held his hand.

As the truck's headlights jerked fitfully across the blackness, he wondered what she was thinking. If it had been daylight, the place might have felt very different; the vistas could be incredible, tens of thousands of miles on either side of unpopulated tundra with areas of thick forestation, low shrubs, moss, lichen, and weeds.

A true and empty wilderness.

Ten thousand years ago, it would have looked very different. The terrain surrounding them would have been made up of vast pastures populated by an incredible density of animals. Herds and super-herds, with a biomass rivaling the biggest cities of modern times. Millions of oxen, bison, horses, and even bigger herbivores, living together, harmonic, symbiotic.

But now there was nothing.

"It's still hard for me to believe," Anastasiya said. "You grew up out here, by yourself. It's like something out of a novel. Such a harsh place for a child."

"I didn't know any different. In the summer, it's actually quite beautiful. Fishing, hiking — it's everything that a little boy could wish for."

"But the winter."

"Yes."

In this part of Siberia there was no sugarcoating the winter. Temperatures that regularly reached −60°, that *averaged* −40°

Celsius. And three months of complete darkness, with winds that could tear the frozen skin from your cheeks.

"My father brought me here when I was two years old. And I was here until I left for high school."

Nikita's sister had gotten out first. As soon as she was old enough, she had left for St. Petersburg, and a world that was not locked in ice. Nikita had followed her lead. Novosibirsk might still have been in Siberia, but with a population of 1.3 million, it was a veritable nation of its own compared to Chersky. The university was first-rate, and Nikita had planned to pursue math and computer modeling, then one day end up in a big city like his sister. He might have had the heart of a scientist in him, but he was not his father. He didn't think, at twenty, that he was willing to give up everything just to save the world.

"But then he came to the dorms four months ago and asked if I would come back."

Anastasiya looked at him. He kept his eyes on the road. She wouldn't understand yet, but it really had been as simple as that. His father had asked him to come back, and he hadn't been able to say no.

"You think I was alone here as a child,

and in some ways it certainly was lonely — but I was a part of something much bigger than myself."

They took another corner in the makeshift road, and through the darkness he could just barely make out the squat building that composed part of the Northeast Science Center. The front lights were on, casting a dull cone through the swirls of snow, and in the center stood a figure draped in a heavy winter coat. The fur-lined hood was up, revealing the man's chiseled, weathered features, above his long, flowing dark beard. He was grinning into the cold.

Sergey Zimov was the strongest man Nikita had ever known. He had built the Northeast Science Center with his bare hands, starting off with little more than a wooden hut in 1980, almost thirty years earlier. Eleven years after that, when the Soviet Union collapsed, his superiors in Moscow had ordered him back to Vladivostok to continue his work. He had refused, instead using whatever resources he could find to store enough food, gasoline, and materials to continue his work through the chaotic governments that followed. Over that time, he'd built the Science Center into a state-of-the-art Arctic research hub, with multiple biological labs, atmospheric data collectors, and tools for geo-engineering. More significant than any of that, thirty miles down the Kolyma River he'd begun working on a dream: a glimpse into both the past and — if they continued what he'd started — the future.

Nikita had no doubt that his father would spend the rest of his life fighting for that dream. He wasn't just the strongest man Nikita knew, but also the most determined. And one of the smartest.

Nikita had never been able to refuse him.

"The family business," Anastasiya said. "You couldn't walk away."

Nikita glanced at her and to his surprise, she didn't look terrified or miserable or angry. She didn't look as if she was ready to turn right around and head back to the little airport, back to the safety and comfort of Novosibirsk.

She looked as if she, too, was coming home.

Chapter Six

Early Spring 1964
CLEARWATER BAY, FLORIDA.
It was one of those rare Saturday afternoons when George Church somehow found himself alone in the house.

Finished with his chores, well ahead of his schoolwork — although at ten years old in the swampy suburbs of a place like Clearwater, that wasn't saying a whole lot. The public middle school Church attended was basically a holding pen, a corral for children. It was simply a place to keep the little hellions all in one place while their parents went on with the business of life.

Certainly, by this age, George's teachers knew about his affinity for numbers. His mother would tell her friends that little George could do math like the wind. But Clearwater wasn't the sort of place where anyone had any expectations. As with most schools in the area, his public school didn't even have a science teacher before seventh grade. And none of George's classmates or friends had any desire to go to school beyond the legally prescribed limits; the idea of college was as foreign as snow.

On the surface, there had been a fair deal of change in George's young life over the past seven years. His mother had replaced his previous stepfather with Dr. Gaylord Church, an educated and worldly pediatrician who had taken an immediate liking to his precocious stepson. The elder Church often carted his stepson with him on trips to conferences in far-flung places, and unlike his classmates, George had become quite cosmopolitan by the beginning of his second decade. He had explored solo the streets of Dubrovnik, Plitvice, Rome, Cuzco, and, with company, Bogotá, Lake Titicaca, Rio di Janeiro, and São Paolo.

He'd also developed a new hobby — going through his stepdad's medical bag to play with the various instruments he found inside. Eventually, he had made his way to a collection of his dad's hypodermic needles. When his stepfather noticed his interest, he taught George how to use one and had George inject him from time to time. George didn't realize until he was much older that his stepfather had been addicted to painkillers, like many doctors of his era.

George could only guess what his friend Charlie might have done with ready access to needles and various classes of opiates; sadly, Charlie was long gone by that time, shipped off to a foster home where he would have to continue his rebellions on his own. In Charlie's place, George had gained a pair of stepbrothers who had come along with Gaylord Church, and two additional siblings as a signing bonus from his third father's first marriage.

Consequently, George had gotten adept at changing diapers, warming bottles, and dodging familial drama. One of the sons of his third father was particularly impressive; the unfortunate teenager had been institutionalized after an accident involving a Frisbee and a high-voltage telephone wire, which had led to a fall down a fire escape, a pair of broken arms, and hands and body covered in burns. Whenever he visited the Churches' house, he reminded George of Frankenstein — arms in matching casts, scars all over his face, and a penchant for never wearing a shirt, the better to reveal the patches of new skin sewn all over his chest.

So, for once, it was a nice surprise to find himself alone in the house at the peak of a

Saturday afternoon. Resting against the couch in his living room, Church had instantly lost himself in that place between awake and asleep. Although his mother had yet to officially diagnose him, Church was narcoleptic as well as dyslexic. At that age, it wasn't something he had entirely come to terms with — at some level, he was aware that people noticed his constant sleepiness, but he'd assumed that everyone was sleepy, but most were better at hiding the problem. Teachers would sometimes throw chalk at him, but it had never really sunk in — *here was another way he was different.* When he wasn't moving, he could instantly fall into a deep sleep. This would happen in the classroom, on the playground, sometimes even in the middle of a conversation. To combat the disorder, Church had developed numerous tricks — bouncing the balls of his feet against the floor, drumming his fingers against any hard object, shifting his head on his neck. But sometimes, he just allowed it to happen.

His mother had assumed his sleepiness was the natural result of a brain that seemed to work twice as fast as everyone else's, but Church doubted that was true. After all, his mind didn't slow when he was asleep — sometimes it seemed to spin even faster. At the moment, his head against the couch pillow, his long legs curled beneath him, he was thinking ahead to the science project that had consumed much of his recent days.

A week earlier, he had been leafing through a science fiction novel in the public library and had stumbled upon a story about giant, man-eating plants. He had immediately decided to try to make one of his own. He had previously learned about Venus flytraps — plants that could trap insects, survive on protein, and were indigenous to the Carolinas — and had begun to study what might make a plant like that grow.

In some old science books, he'd found a chemical that could significantly exaggerate the size of bean sprouts, and he intended to apply them to a crop of flytraps he was growing in the front yard. His goal wasn't to terrorize the neighborhood — though that would yield a certain satisfaction — but simply to prove that it could be done. *A new mental exercise, like blowing up plastic cowboys with fireworks.*

But today, his thoughts of massive, toothy greenery were interrupted by a sudden knocking that cut through the fog of his sleep. He was halfway off the couch before he realized that someone was at the front door.

Crossing the living room and opening the door, he was surprised to find one of his neighbors standing on the porch — a middle-aged man in jeans and work boots, with a baseball hat pulled down over his eyes.

"George," the man said by way of a greeting. His southern accent was so heavy, his syllables extended themselves so long, that Church had to employ his physical tricks to keep his focus.

"Can I help you, sir?"

"Just wanted to congratulate you," the man replied. "I see that you're finally taking care of that jungle behind your house."

Church raised his eyebrows. One of his main chores from age ten was yard maintenance. The house Gaylord Church had moved them into was set on about an acre of land. Although the neighbors had mostly beautiful, well-manicured lawns that looked like golf courses, Church's yard was covered in an angry species of weed called sand spurs. Church's mother paid him a penny for every plant he ripped out, but clearing the lawn of the vicious spurs was a Sisyphean task.

At some point in the past few weeks, he'd come up with the idea to go after the spurs with science. He'd gotten hold of a magnifying glass and discovered that, by focusing the intense Florida sun at the spurs, he could burn them to the root. That very morning, he had cleared a good quarter of their backyard.

"Doing my best, sir," Church said, pleased with himself.

The man nodded, then turned to leave. Over his shoulder, he said, "Oh, and nice fire."

Church didn't show any emotion until the man had stepped away from the porch. Then he slammed the door shut, turned, and ran across the living room toward the back of the house. Even before he reached the back door, he could smell the smoke. Throwing open the door, he immediately saw that he'd set half the yard on fire. The flames were high in the air, spreading outward from the gnarled trees toward the area that separated his lawn from the neighboring community.

Church went right for the garden hose curled beneath the back garage. Luckily, he had developed an extensive knowledge of compression and water pressure, and knew exactly how to turn his regular hose into something that resembled a crop duster.

Even so, it took a good forty minutes before he'd gained control of the conflagration. He was finishing rolling the hose back into place when he heard his mother's car pull into the front driveway. He made it back inside just as she reached the front porch — then caught sight of himself reflected in a picture frame above the couch. His face was covered in soot, burnt spurs sticking out from his hair like a demented crown. He dived into the bathroom as his mother got her keys into the door, then jammed his head under the faucet.

As his mother entered the living room, he was still shaking water out of the wild locks of his hair. His mother gave him a curious look, but didn't say anything. Eventually, Church knew, she was going to glance out the back door and see their scorched trees, but for the moment he was in the clear.

Which was a good thing, because his mother had a surprise for him.

"George, you're going to miss school next week."

"Are we going somewhere?"

Missing school was far from a big deal; the most important thing on his schedule for the week was a violent game of dodge ball in PE.

"Not so much where," his mother responded, "as when."

And as the smoke from the backyard fire

rose past the windowpanes behind him, she opened an envelope containing a pair of two-dollar admission tickets to the 1964–65 World's Fair.

In real life, time travel wasn't as simple a process as comic books and sci-fi B movies had led Church to believe. It didn't involve Plexiglas tubes and blinking neon lights, faraday cages spitting electricity, control panels leaking swirls of colored smoke.

Instead, time travel began with a three-day drive up Interstate One, trapped in the passenger seat of his mother's '62 Buick. Along the way, Church had hung his head as far out the side window as decorum and the proximity to surrounding traffic allowed. He'd suffered from extreme motion sickness for as long as he could remember, so it was a good thing that time travel also included frequent stops at the homes of various cousins, uncles, and aunts up and down the eastern seaboard, a copious amount of fast food from highway restaurants, and a four-hour traffic jam on the Queensboro Bridge — after which the chaos of the vastly overwhelmed parking lot at Flushing Meadows seemed almost civilized.

Even so, hours after leaving the parking lot, Church was still fighting the tail end of an almost continuous bout of nausea as he wriggled out from underneath a polished safety bar, and away from the caterpillar of moving chairs that had taken his mother and him on a tour of Futurama, an exhibit located within the General Motors Pavilion. Riding in his chair, Church had just gotten a glimpse into life thirty years into the future, with inventions that included brightly colored concept cars — sleek, slick vehicles with bubbles of glass for cockpits and fins that looked as if they had been pulled from the tails of a jet — moving sidewalks, even moon colonies and underwater hotels.

No amount of motion sickness could have dampened Church's excitement. The 1964–65 World's Fair had surpassed what he'd imagined during the long trip up the East Coast, from the moment he'd entered the fairgrounds and caught sight of the massive, 120-foot-tall globe, the Unisphere. Made of crisscrossing curved bars twisted into a perfect sphere, the Unisphere was the focal point at the end of a long avenue lined with flags from every country on the planet. The fair's theme was *"Peace through understanding,"* and it was dedicated to *"Man's Achievement on a Shrinking Globe in an Expanding Universe."*

This was the future, as imagined by a collaboration of automobile, oil, manufacturing, entertainment, and even computer companies. They had hired the world's most prominent architects to build impressive pavilions that filled Flushing Meadows from one end to the other. And within some of those pavilions, many visitors had their first encounters with the new world of computing.

To Church, everything at the fair seemed shiny and brilliant. In the IBM Pavilion, he'd had the chance to see a mainframe computer up close. In the Ford Motor Pavilion, he'd taken a ride in a quad of Ford convertibles along a skyway whose scenery traced the history of life on Earth, from the dinosaurs through the present to an imagined future not unlike the one depicted by their competitors at GM in Futurama.

Walking next to his mother past a domed car that looked more like a grounded jet airplane, with fins supporting what looked to be rocket tubes and wheels that might have twisted flat for takeoff, Church felt his mom's hand on his shoulder. He knew she could feel that he was trembling, and he half expected her to diagnose him with some new neurosis or disease. He wanted to put her at ease, but excited as he was, he also felt a strange anxiety growing in the pit of his stomach.

"It's not real," Church finally said. "I mean, I know it's not supposed to be real. It's supposed to be what's coming. What's possible."

His mother looked at him. Church shrugged his wide shoulders. He was already twice as tall as most of his classmates, and his growth showed no signs of slowing.

"This future — this is where we ought to live," he said. "But we don't. We can't."

At the Disney Pavilion, he'd seen an animatronic President Lincoln. At first glance, it had seemed so impressive, a machine of gears and levers that looked and talked like a person. But when he'd leaned in close, he'd seen that the paint was faded and the screws' heads stripped. He compared that worn robot to the ride he'd just taken, to the imagined cities of the future, deep underwater, in space, on the moon. These were the imagined results of the atomic age come to life, the harnessing of energies well beyond present understanding. Everything at the fair was so incredibly smooth, bright, and shiny compared with the present, which felt so broken down. Unwieldy. Rough and busted.

"None of this is going to be real, unless

people make it real."

Church knew he wasn't finding the right words. At ten years old, his vocabulary hadn't yet caught up with his imagination and capacity to think. But he also knew that, compared with what he'd just seen, the present was so — *boring.* He didn't want to be *here,* he didn't belong *here.*

Later in life, he would return to this moment as the instant when he first started to think of himself as a sort of time traveler. Deep down, he started to believe that he was from the far future, and had somehow been left in the past. It was his task in life to try to get back, to try to shift the world to where he had once been.

From the look on his mother's face, he could tell that she understood.

"So you'll make it real," she said.

She was going to help him any way she could. But, still, he knew, it would be up to him. If he wanted to reach the future, he would have to figure out a way to write it himself.

CHAPTER SEVEN

December 10, 1984. Twenty years later.
ALTA SKI RESORT, WASATCH MOUNTAINS, UTAH.

At ten thousand feet above sea level, the air was so thin that it was like breathing through a straw. The temperature was a brisk forty-two degrees, and the sky above a dangerous shade of gunmetal gray.

Midmorning, the snow had begun to fall again, the heavy flakes swirling down from the low cloud cover, diminishing visibility to a paltry few yards. If the weather reports were accurate, in a resort town that lived and died by the accuracy of its meteorologists, the area was facing its second blizzard in three days. Many of the roads that traversed the sixty-mile passage back to the Salt Lake City airport were already closed, and the long ride up the triple chairlift from base level had been like ascending through the highly charged outer atmosphere of one of Jupiter's moons.

At the moment, Church couldn't have cared less about the forecast or the visibility. With four feet of fresh powder, the runs at the summit were pristine and untouched. His own skis barely left tracks as he cut a tight line down the center of the mountain. Each twist of his hips sent showers of snow toward the tree line on either side of the narrow trail, and the mist from his breath crystallized in lavish patterns against the bluish glaze of his goggles.

A towering six foot five by his late twenties, with wide shoulders and long limbs, Church had grown into an avid athlete. Skiing was just one of his many physical hobbies. Growing up in Florida, he hadn't expected to develop an affinity for a cold-weather sport. The only skis he'd ever been on had been in water, back when he was a toddler and still living on the air force base of father number one. But when his mother had saved him from the unexceptional school system in Clearwater by packing him off at age fourteen to Phillips Andover boarding school in Massachusetts, he'd opened his mind to an abundance of new experiences.

Compared to a childhood spent in depressed Florida public schools, Andover had been like the World's Fair all over again. Everything was shiny and new, and students had access to nearly limitless resources. His classmates were either smart enough to keep up with him or understanding enough to get out of his way. Immersing himself in exploring a wide range of subjects, he'd rapidly expanded his interests: biology, chemistry, higher math. After finding an ignored computer in the basement of the science building, he'd taught himself to program and done a bit of hacking just to see if he could.

By the end of his second year, his teachers recognized that he was different, and, eventually, they gave him keys to the various labs, so he could conduct his independent projects on his own.

At Andover, he first discovered the world of genetics. He was fascinated by the fact that every living cell had within it a genomic code — a double helix of DNA made up of bases of chemical molecules attached together like rungs on a ladder. And that ladder comprised coding for everything, from eye color to the length of fingers and toes. At the time, the study of genetics was in its infancy, and it was the dirty stepchild of the biological field. People who wanted to change the world didn't go into genetics;

they found other avenues for making significant contributions.

Church bent low on his knees as he came into a tight turn, his ski poles bouncing at his sides. He could no longer hear the skis of the colleagues he had set off with from the base that morning; a quick glance over his shoulder gave him no new information. They could have been a dozen feet behind him or he could have lost them at the last fork. For all he knew, they had stopped at one of the upper mountain cafés and were enjoying cocoa in front of a fire.

Over the past decade, Church had lost the ability to take it slow. Trading Andover for Duke University, he had been in a rush to get through college. He'd skipped most freshman and sophomore classes and headed directly into the advanced sciences. At Duke, he had taken additional classes and done extra work for no credit, especially whenever there was a lab component. While earning his degree in chemistry and zoology, he'd continued to develop his knowledge of genetics on the side. During the summer between his first and second year, he'd applied to graduate schools, earning admission to several schools, including Harvard, but had chosen to remain at Duke.

Church's focus became crystallography:

The field seemed a natural fusion of all of his interests — math, computers, chemistry, and biology. He had especially taken to the professor in charge of his crystallography lab, Sung Hou Kim, who had just arrived from his own postdoctorate at MIT. When Church had gone into the young professor's lab for his initial sophomore-year work interview, Kim had been working on a large brass molecule construction kit, attacking the faux atoms with a little steel wrench, smiling. Church had loved the physicality of the exercise, and had decided then and there that it was where he belonged.

Working in Kim's lab became an obsession, and Church stopped attending his other classes. He taught himself to become an expert in the study of RNA, the messenger molecules that carry out tasks inside the cell, enabling the development of trait-generating proteins. Soon, Church was spending more than one hundred hours a week in Kim's lab.

It wasn't until he received a letter in the mail from the Duke administration that he realized he had failed out of the Ph.D. program. Although he'd published five notable papers in that first year, all that mattered to the university was the fact that he hadn't attended any of his necessary course work and was now receiving failing grades in two of his classes (which he had already aced as an undergrad).

Years later, he still had that letter, as a reminder to himself of the first time he'd nearly derailed his entire career:

> Dear Mr. Church
>
> Examination of your record for the past semester reveals that you earned a grade of F in a course in your major field. Earning a grade of F in the major field occasions withdrawal from a degree program (see page 58, Bulletin of Duke University, The Graduate School). Cosequently, you are no longer a candidate for the Doctor of Philosophy degree in the department of Biochemistry at Duke University.
>
> We regret that this action is necessary and hope that whatever problems or circumstances may have contributed to your lack of success in pursuing your chosen field at Duke will not keep you from successful pursuit of a productive career. . . .

The dismissal had come as a major shock; Church had completed his undergraduate degree in only two years, and had gotten kicked out of graduate school in less than half that time. It was a quick and painful lesson in the culture of academia; fulfilling requirements was somehow valued over whatever original research or work one might be doing. Church considered giving up graduate school entirely and focusing on lab work, but Dr. Kim rescued him from his own shortsightedness. Kim asked him, did he want to be a technician, or did he want the autonomy that would better enable extreme innovation? If he continued on the road he was going down, he would never have any control. He would always be at the mercy of someone else. Church took his mentor's advice.

On the next turn of the mountain slope, Church was moving so fast he felt his outer ski lose touch with the snow. For a brief second he thought he might skid out of control. Then he countered with his entire weight, came up centered, and continued rocketing down the slope.

Getting into a Harvard Ph.D. program hadn't felt like a life-changing event at the time. Compared to Duke, Church's Ph.D. work had been free of drama, and after graduating he'd spent six months at a biotech company in Boston, then a few months out West. After that, he'd gone back to the Northeast, where he'd fielded offers and almost ended up at Yale, before he'd learned how little they were willing to pay to set up his lab. Then he'd given a talk that, oddly, turned into an interview at Livermore, where they were building nuclear bombs. He'd had to enter the facility through a sequence of three barbed-wire fences and he'd quickly decided that wasn't for him, either.

Luckily, a friend at Harvard, Gary Ruvkun, told him about an opening in the genetics department of Harvard Medical School. Although Church was a little out of cycle and the department had just finished its hiring process, he'd applied anyway. At his interview, he pitched his Ph.D. thesis, a dissertation about a new method of analyzing genetic enhancers that were involved in DNA-binding. It was just sexy enough to get him the post, but what he truly wanted to work on was sequencing — coming up with ways to read genetic material fast, better, and cheaper.

At the time, the art of reading genomes was just being developed, although James D. Watson and Francis Crick had discovered and defined the double helix structure of DNA as far back as 1953. It was now well known that genetic material is formed by billions of organic molecules called nucleotides (the basic units of nucleic acid that make up DNA and RNA). These are arranged in sequences that create the genome, which is carried by every living cell in every living organism, and in which all traits are coded.

But isolating and "reading" those sequences was still extremely difficult. Frederick Sanger had only sequenced the first full genome — that of a simple virus — as recently as 1977. Any high school biology teacher could explain that DNA was made up of four chemical building blocks — adenine, cytosine, guanine, and thymine (ACGT) — that were paired together and attached by hydrogen into a molecule that resembled the steps of a twisting ladder, and that this genetic material, duplicated in every cell, was essentially the instruction booklet behind life itself. But making sense of those instructions, and understanding which sequence of ACGT gave rise to what specific characteristic, was exceedingly difficult.

The process of isolating and reading genomic codes originally involved giant devices with big chunks of paper covered in genetic material that were dipped in vast tanks of kerosene as coolant. A scientist would sit in a room surrounded by fire extinguishers as he or she ran six thousand volts of electricity through the paper, hoping the whole thing didn't explode. The process that came after that one was a bit more fine-tuned, and also less flammable, but researchers still had to lug clunky, meter-long pieces of glass around the lab. Then came an automated technique, called capillary electrophoresis, which involved rooms full of giant machines that cost millions of dollars.

As part of Church's Ph.D., he'd developed a way to multiplex the process of reading genomes, essentially speeding up the technique and making it cheaper by sequencing multiple strands at once. He'd likened it to moving from a swamp to a village — his goal was to eventually get to a full-scale city. From the beginning, his multiplex technique garnered immense interest from all across academia. A process that had previously taken months, or longer, could now be done in days.

At his Harvard office, Church had received a phone call from the Department of Energy just weeks before traveling to Utah, asking him to be part of a unique project. He had assumed that the venture would be connected to his work on multiplexing, but to his surprise, the government had something else entirely on its mind. The DOE was bringing together the top scientists in genetics in order to try to figure out the mutation rates among populations who had been living downwind from Hiroshima when America dropped the first atomic bomb.

Church had been the youngest scientist on the invitation list, which had included seventeen other prestigious names. Even so, he had been less than enthusiastic about the DOE request. He hadn't gone into genetics to deal with the aftereffects of weapons of war as a top consideration. But David Smith of the DOE and the rest of the brass there had made several convincing arguments that appealed to Church: The meeting would take place at one of the premier ski resorts in the country, Sunday, December 9, to Thursday, December 13, 1984, and it would be paid for by the Department of Energy and the International Commission for Protection Against Environmental Mutagens and Carcinogens. There was fresh powder in the forecast, and some amazing scientists whom Church really wanted to meet had already accepted the invitation. Being a nerd, he found the latter as compelling as the former.

Almost immediately after the scientists convened at Alta, they'd come to the conclusion that there was no reliable way of measuring the cellular effects on people living downwind from chemical or nuclear explosions. Trapped in place by multiple blizzards, they had instead decided to make the most of their time by hitting the slopes and then meeting in hot tubs and at the main restaurant for long bull sessions on anything that struck their fancy.

Church wasn't sure by what magic their conversation shifted to significant ideas, but he assumed it was just the natural result of isolating so many brilliant people together for five days. Late that first night, the idea came up of reading the entire human genome.

By determining all the sequences of A,C,G,T, they would, in effect, figure out what made a person a person, down to the cellular level. It was as audacious a project as the original moonshot, which had led to men walking on the lunar surface. But instead of looking outward, toward space, theirs would be an adventure inward, toward the center of humankind.

The first question that had arisen was how much such a project might cost. At the time, their estimate was a dollar for every base pair of genes, three billion dollars in total. Church was the only one there with a computer, aTRS80 model 100 — a 1.5 kg, pre-laptop-era laptop — and he started doing reality checks on the wild scenarios that he and others were considering.

Almost immediately, the Department of Energy began writing checks. One thing the government of the eighties was good at was writing checks in support of science. For Church that would mean that, as soon as he got back to Boston, he would scramble to put together grant proposals. A running joke in his lab in those early days was that he was the only person at Harvard who didn't know how to properly ask for enough money. But the Human Genome Project (HGP), as the DNA-mapping effort would soon become known, would change that. Soon there would be considerable amounts of money going around, and even an inexperienced grant writer with some odd ideas could find a way to get some water from the spigot.

The Human Genome Project would hit the biological sciences like a lightning bolt. But in its first conception at the lodge in 1984, it was just a way to turn a blizzardy week with a peculiar goal into something productive.

Church hit a tuft of snow head-on, then shifted his weight to take a narrow curve in the trail. Maybe it was the shadowy gray light, barely peeking through the low clouds, maybe it was the snow that was now gusting horizontally through the air, but he never even saw the tree stump, jutting up from the edge of the slope. His first ski managed to clear the frozen wood, but his back ski collided with an ugly jolt. Church tumbled forward, his long frame jackknifing over the tips of his skis.

Luckily for him, the fresh powder was soft enough, even for a headfirst landing. He hit and kept on rolling, sending up a massive cloud of snow.

By the time he finally heaved to a stop and started to crawl his way out from under the snow, thick clumps hanging from his thick beard like Christmas tree ornaments, two of his colleagues had caught up, shushing to a stop just a few feet away. David Smith from the DOE was in front, yanking his goggles up to look George over. Then he smiled.

"For a minute there, I got a little bit worried about the future of DNA sequencing."

Church laughed.

He always had a knack for making an impression.

Chapter Eight

December 14, 1979. Five years earlier.
WEEKS BRIDGE, HARVARD UNIVERSITY, CAMBRIDGE, MASSACHUSETTS.

At ten minutes past midnight, Chao-ting Wu leaned out over the dark water of the Charles River, her elbows heavy against the stone railing of the seventy-year-old bridge, and wondered to herself how someone so incredibly smart could do something so incredibly disappointing.

Ting, as she had been called since childhood, could feel Church standing awkwardly next to her, pretending to look out over the water as she was — but he wasn't fooling her at all. She knew exactly how his mind worked. He wasn't contemplating the way the moon danced across the glassy surface of the river, the refracted light made even brighter still by the template of a pre-crystallizing liquid, H_2O kept in a state just a few degrees above freezing. Nor was he counting the chunks of ice that had come loose from the snowy banks on either side, floating in aimless circles, buoyed by some incalculable logic of water current, wind, and temperature.

Ting doubted Church even saw the river, the ice, or the moon. She knew he was too busy going over the details of the evening — from the minute he had picked her up at her apartment to attend a fellow student's party to the moment when he had finally taken her by the hand. No doubt, now he was contemplating what he was supposed to do next. Ting was glad that so far, those ruminations had led only to his standing in awkward silence, just a little too close, as they leaned together over the Charles, while the fingers of his free hand absentmindedly pulled at the end of his thick, ever-present beard.

Ting could honestly say she had never expected this moment. She had met Church shortly after he had entered graduate school at Harvard. They had taken a course together on chromatin structure, and Church had sat across from her. He was tall and so, so quiet, but every time he opened his mouth, whatever came out was strange, smart, and creative. Even so, people had begun referring to him as "three-words-

George," because that was as much as anyone ever seemed to get out of him.

Ting had taken an instant liking to him. From the very first week, they had spent many hours talking science — always science. They would meet in one of the Harvard libraries to exchange ideas or they would take walks along the Charles River and around Harvard Yard. There was so much that was bold and archaic and looming about the place, from the stone architecture of Widener Library to the brick and antique glass of the old dorms like Thayer and Wigglesworth. In contrast, their conversations were so new and modern: crystallography, life sciences, the nascent field of recombinant DNA.

Unlike Church, Ting had fallen into biology quite by accident. She had come to Harvard to study math, but while working as a dishwasher in a biology lab to help support herself through college, she had suddenly felt at home. The simple beauty of the natural world held so much in common with the purity of numbers — she and Church could go on and on about the subject. Ting loved having a male friend with whom she could talk just that sort of deep science.

So it had come as a complete shock to her

when, one day in the library, Church had turned to her and asked if perhaps she might think about having their interactions become more social. It was such a Church way to ask — and her response had been immediate disappointment. She had found this great friend, and suddenly he wanted more.

From then on, she had avoided him. If he came near her in the halls, she ran up the nearest set of stairs. Months passed, and her feeling of disappointment stayed with her. She would have talked science with him any time; indeed, their conversations were unlike any either of them could have with anyone else, but that was as far as she wanted it to go.

But just a few days earlier, a friend of hers had come back from a grad school party and had told her she'd seen Church turning down numerous women who wanted to dance with him. Ting had felt instantly responsible, and she decided she needed to tell Church it was time for him to move on.

His response had surprised her: He said he would wait five years for her. Ting knew that this was not an arbitrary declaration. Church's willpower and determination were well-known. Just before graduate school, he had taken part in an MIT nutrition study and had volunteered to subsist on a bowl of cornstarch and a plastic test tube full of amino acids for forty-five straight days. Worse yet, he'd had to keep complete track of whatever went into his body and whatever came out. Which meant once a day, he'd had to return to the lab with bags full of the end result of a diet of cornstarch and amino acids. Nobody else in the study had lasted half the prescribed time period, but Church had never even considered giving up.

Five years pining for her would've been a walk in the park. So Ting had made him a deal: one date, and he would see how incompatible they were outside a library.

It wasn't until just minutes ago — after the party, the walk to Weeks Bridge, the touch of his hand in hers — that she had begun to realize something surprising: She had been wrong, and he had been right.

December 14, 1990

CAMBRIDGE CITY HALL,
795 MASSACHUSETTS AVENUE,
CAMBRIDGE, MASSACHUSETTS.

Eleven years to the day from that first date, Ting was still holding George Church's hand. They had traded the view from Weeks Bridge for a court-appointed justice of the peace, who had stepped out from behind his oversized mahogany desk long enough to read a few lines from the marriage license they had just finished signing.

Around them, the walls, staircases, and much of the ceiling were trimmed in ornate wood molding and the floors were mostly marble. From the outside, the building was impressive, three stories of Romanesque masonry, complete with arched windows and a lofty bell tower casting shadows down the bustle of Massachusetts Avenue. But inside, despite the wood and marble, there was no mistaking the feeling that this was a hundred-year-old government office.

All of which was perfectly agreeable to Ting and Church. They had reached their decision to get married with a similar lack of fanfare; they had simply decided it was time, since they were ready to start trying to have children. There had never been any question of having a traditional wedding, or even of telling most of the people they knew that they were getting hitched. In fact, they had biked to city hall. The only concession to the formality of the moment Ting had made was that, for once, she had actually worn a skirt. And this was balanced by the fact that the witness they'd chosen for the proceeding was the clerk from the liquor license window across the hall.

Over the past eleven years, the main beats of their relationship hadn't changed; they were both true scientists, and when they were together, they talked science. In social settings, Ting had helped George come out of his shell. More often than not, he now spoke in full sentences, and had endeared himself to all of Ting's friends. Everyone who knew Church adored him; he was physically big, overwhelmingly bearded, and mentally astounding, but he wasn't at all intimidating.

At Harvard, Church's career had taken off like a rocket. From his work in multiplex genetic sequencing to his helping to found the Human Genome Project, he had gathered an immense amount of momentum behind him. His lab was growing at an astonishing rate, as was his standing at the university. He had become a unique voice in science, and his lab was filled with unique minds. The letter grades of his student and lab staff meant less to him than their creativity and willingness to break barriers. Creating genetic technology was only going to get faster, smarter, and cheaper. His goal was to build a lab that always saw two steps ahead of others.

Ting's journey had not begun quite so auspiciously. She had chosen to work on research involving the genetics of fruit flies through the mechanics of their chromosomes, starting with her early work on telomeres and now focusing on how chromosomes interact with each other. As in all living creatures, the fruit fly's genome is made up of long chains of DNA — the building blocks of life that code for everything from the length of a fruit fly's wings to the color of its body. These long chains created separate chromosomes. Whereas humans' genetic material contains twenty-three pairs of chromosomes, for a total of forty-six individual chromosomes, a fruit fly has four pairs. Telomeres are specific, repeated DNA sequences that appear at the end of each chromosome in humans, fruit flies, and all living creatures. Telomeres protect those chromosomes from degradation during replication, when errors can occur that can give rise to disease. Essentially, telomeres act as chemical buffers, or bumpers, to help keep genetic material intact. Even so, over time, telomeres shorten and eventually degrade, a process many scientists believe is a cause of aging.

Ting's scientific research was complex and intense, which she enjoyed. But the trouble for her started almost immediately, as she went to look for a faculty position in the

Department of Genetics, where Church was already a faculty member. Although she and Church were not yet married at the time, and had told very few people that they were even in a serious relationship, the head of the department had made it clear that she was not to be accepted; he believed it would be a conflict of interest for her to be in the same department as Church.

To Ting, it was infuriating; her life with George was personal, and it made no sense that anyone would, in these modern times, still argue that her personal life could be used to dictate her professional capacity. Relationships, marriage — like skin color — were a personal feature, and should not be used as a measure of a person's worth or ability.

In fact, she and Church rarely saw any need to share their private life with anyone. When Ting had become pregnant a short time after their marriage, many of her friends had even expressed concern about how difficult it would be to raise a kid on her own. "But George and I are married, have been together for eleven years," she had explained.

Since there was no official university policy against the spouses of scientists applying to the same department as their husbands, Ting stayed the course and, despite his warnings, applied to the genetics department, but covered her bases by also applying to every other department that fit her interests. She was accepted to the Department of Anatomy, shortly after which she went on maternity leave.

It wasn't until she was about to return from leave that she discovered that the Department of Anatomy had been dissolved, with parts subsumed into various other departments. Since she was a geneticist, she was advised to request that her transfer be to the Genetics Department. This, however, would put the same director who had refused her entrance in charge of her life. She immediately set up a meeting with the man, but once they were face-to-face, he made it very clear that he did not want her to stay. He had a personal policy against the hiring of spouses of tenure track professors — under no circumstances would she ever earn her own tenure under him. He suggested she give up her faculty position to work as an assistant in another faculty member's laboratory, as then he would be comfortable with her being a member of his department.

Ting was angered and insulted. She had experienced racism because of her background, she had lived through sexism because of her gender, and now she was learning that her marriage to one of the most promising men in all of biology was actually going to be an obstacle to her career.

Church was equally shocked and wanted to help, but the situation made outright action on his part difficult. Instead, he and Ting worked as a team, with her on the front lines and Church as a sounding board, providing whatever insights he could. But they had been trained as scientists, and this was unfamiliar territory. They were discreet. Few knew what was happening, and those who sensed something was amiss wondered whether the problem lay with Ting, some even speculating that she was a mediocre scientist who had gained access to the department only because she was Church's wife, thus justifying the department chair's opinion. The irony of the situation did not escape Church and Ting, as they knew the difficulties she faced. She was assigned a tiny lab that amounted to a fraction of the space normally offered to faculty of her position, and students were not allowed to work for her.

Every morning, Ting and Church walked to work together; he would turn right and head into a state-of-the-art, well-financed lab full of brilliant postdocs, and she would turn left and head into her closet of a lab, which had been diminished at one point to a single other researcher. Church found the situation unbearable.

When the time approached for her to put up her name for tenure, Ting and Church knew that the damage to her career was insurmountable. Against all odds, she had established herself as a pioneer in her field, with international renown, but there was no denying that she did not have anywhere near the number of published papers one would normally expect of someone of her caliber and position. She simply had not been afforded a proper lab or the students working with her that were necessary to build a publishing legacy. She knew what she needed to do — there wasn't a choice. Before a panel of senior faculty, she requested that her tenure packet include a letter explaining the circumstances under which she had worked. Her chair did not agree, and the other faculty in the room did little better than watch the event unfold. Nevertheless, Ting had met her goal, spoken up for transparency and fairness.

Her tenure evaluation was one of the strangest that the Harvard Medical School had ever conducted. She heard, after the fact, that the senior faculty members had been so confused by her CV, some of them asked if there was some sort of lawsuit going on that they didn't know about. The head of the department didn't explain, he simply pushed them to refuse her request.

She was not surprised by the outcome. She turned her attention to securing the immediate functionality of her lab and approached a newly minted dean, who was appalled at the situation and on the spot got Ting interim funding. In the meantime, Ting and Church began looking for new jobs at other universities. If Harvard didn't want Ting, Ting and Church didn't need to stay at Harvard. However, word of the odd state of her CV reached as high as Larry Summers, the president of Harvard at the time.

Before long, the dean called again, stating that she would like to take the stunning, highly unusual step of rerunning Ting's tenure process, without the knowledge of the department chair. At that point, Ting and Church were already trying to decide between offers at the University of Washington in Seattle, Washington University in St. Louis, and Boston University, and were feeling excited by the prospect of leaving for more friendly surroundings. Assured that the restarted tenure process would not be completed before she and Church would have signed new contracts elsewhere and that, therefore, there would be no complications with their departure from Harvard, Ting agreed; she liked the new dean and understood her need to take corrective action. It was the right thing to do. Then Ting got the word — her tenure had been granted. It would be the last tenure Larry Summers approved as president of Harvard. Furthermore, the head of the department who had opposed her would soon be resigning.

The speed and strength of the confirmation of her scholarship after seventeen years of academic inequity were completely unexpected, and that it would come from outside her department spoke volumes to Ting. Tenure was, at this point, just an administrative label. What mattered much more was that, outside the walls of Harvard, she was recognized as a scientist in her own right.

Ting was now faced with a choice. They could move Church's and her laboratories to a university that did not have issues with their marriage, although that would disrupt the lives of their trainees, numbering in the fifties for Church at that time. Or they could continue to work at Harvard Medical

School, thereby sparing their trainees the life disruptions that can so easily derail aspiring scientists. Ting steeled herself to the reality that she would be staying on at Harvard Medical School.

All in all, it had been a traumatic lesson in the politics of science. Enormous ventures like the Human Genome Project (HGP) could change the direction of where the money would flow, but politics would influence how and where scientists lived and worked. Lines of authority would sometimes be arbitrary; and good and bad people would make decisions that influenced scientists' daily lives.

With the barriers lifted from her lab, Ting did right her ship. In fact, she went on to get two coveted "high-risk, high-reward" National Institutes of Health Director's Awards totaling almost $10 million to pioneer new technologies that have enabled some of the highest-resolution images of the genome to date, and to pursue transformative ideas for combating disease. She would wonder from time to time what more she might have accomplished had she not spent seventeen years fighting for equality. Mostly, though, she planned to recoup those lost years by living, and working, seventeen years longer.

Going forward, Church was determined to insulate his young charges from university politics as much as he could. The battles that Church and his lab were going to fight weren't going to be with petty administrators and political tribunals. He was taking aim directly at the parameters of biology itself.

There was only one rule that he had set, as law: Nothing was impossible.

Hand in hand with Church, in front of a justice of the peace, they had only begun to get an inkling of how differently the world would treat them, as a couple, but they were confident that they would fight as a team to push science itself out of its comfort zone and into a future that sometimes only they could see.

■ ■ ■ ■

Part Two

■ ■ ■ ■

A poet sees a flower and can go on and on about how beautiful the colors are. But what the poet doesn't see is the xylem and the phloem and the pollen and the thousands of generations of breeding and the billions of years before that. All of that is only available to the scientists.

— GEORGE M. CHURCH

Every cell in our body, whether it's a bacterial cell or a human cell, has a genome. You can extract that genome — it's kind of like a linear tape — and you can read it by a variety of methods. Similarly, like a string of letters that you can read, you can also change it. You can write, you can edit it, and then you can put it back in the cell.

— GEORGE M. CHURCH

Chapter Nine

Early Fall 2008

77 AVENUE LOUIS PASTEUR, BOSTON.

Sometimes, it's the strange questions that keep you up at night.

Church leaned back in his chair, his long legs tucked beneath the desk in the middle of his stark, brightly lit office, nestled deep in a corner of his second-floor laboratory. His right hand was still resting on the phone in front of him, long after he'd hung up, his feet bouncing against the carpet beneath the desk in the self-taught routine he used to keep himself awake.

He wasn't sure how long he had been sitting there, staring at the dormant phone, after having spoken to a journalist who had posed an astonishing question. He could tell by the dark sliver stretching across the bottom of the drawn window shades on the other side of the concise, ten-by-ten space that the afternoon had shifted to early evening. But in New England, in the fall, he couldn't be much more accurate than that. Nor would it have helped to open the door behind him and peer out into his lab. The young scientists — best guess, now numbering in the seventies — who called the Church Lab home made their own schedules. And most of them had little use for watches, clocks, daylight — really, anything beyond the sterile, cinder-block walls of the Harvard Med School New Research building. At midnight on any Wednesday, there might be twenty-five postdocs huddled around the various Plexiglas sterile hoods lining the twists and turns that separated Church's office from the pair of elevators leading down to the building's lobby.

Church turned his attention back to the phone conversation that he was still contemplating, hours after it was over. He often packed up and headed home early, to see Ting and their daughter, Marie, who had recently turned seventeen. Of course, Marie understood when her parents' hours were erratic. She had basically grown up in a laboratory, raised by scientists who saw her as the most wonderful experiment they'd ever conducted. For her second-birthday party, Ting and Church had lined up Petri dishes for all the kids to grow beans, with varying amounts of vitamins, to see the difference the nutrients made in the plants' early life.

Around that same time, Church had been experimenting with alginate/plaster casting. After noticing that his daughter often fell asleep gripping his thumb, he had used one of Marie's sippy cups as a cast to create a synthetic version of his digit, down to the nail; then, when he traveled for work, a part of him always stayed behind to comfort her and help her sleep. On her third birthday, they'd made an alginate mold of Marie's face. For her fifth-birthday party, they'd made flashlights out of LEDs, along with breadboard circuits.

To the other kids in Marie's classes, her father was a sort of mad scientist, complete with crazy beard and hair. She, too, knew her parents were special. Almost every night, the dinner table conversation reached Ph.D.-level sophistication. In the debates, Church and Ting would switch sides so fast that Marie had trouble keeping up. Sometimes, her father was the one with the outrageous ideas, but often her mother was as well. They took turns pushing each other higher and catching each other's mistakes. From time to time, Marie would join in and push them both or catch them on inconsistencies.

Although Marie and Ting wouldn't find it unusual that he was still at the lab that night, Church wondered what they'd make of the phone call — Ting, especially, since her understanding of the biology that underlay the conversation rivaled Church's own. She'd have known where Church's mind would naturally go. Would she accelerate with him up into that stratosphere or yank them both back down to Earth?

Church rose from his desk and turned toward the bookshelf that lined most of the back wall of his office. It took him a few minutes to find the handful of zoology textbooks he'd kept from his college years back at Duke. Most of them were dog-eared, the covers faded, and some of the information inside was surely outdated. He retrieved a couple of the more basic tomes and spread them out by the telephone, open to equivalent sections, on a specific species of animal.

He sat back down, leafing through the various photos in the textbook, while reviewing the phone conversation again in his head.

It wasn't every day that a journalist called to ask Church to discuss performing a miracle — though it wasn't as rare as one might think. In the past decade and a half, the Church Lab and the brilliant young scientists he had gathered there had pushed the boundaries of genetics, working on projects as diverse as genetically altered mosquitoes to combat malaria and bacteria edited at the cellular level to create powerful new materials. Church himself had authored hundreds of groundbreaking papers in all of the most prestigious scientific journals, and he had been awarded more than sixty patents covering many of his achievements. He'd cofounded more than a dozen companies and changed the practice of both genetic sequencing and genetic engineering.

In theory, the majority of the projects under the Church Lab's wide umbrella had to do with understanding and curing disease, but the ability to sequence and manipulate DNA — the building block of all life on Earth — had limitless applications. Church had been one of the earliest innovators to understand that genetic engineering could create changes in physical matter and physical life that could seem miraculous, given enough time and money.

The Human Genome Project (HGP), which Church had helped originate, had been completed in 2003, in less than the fifteen years that had been projected, at the cost of around three billion dollars — as the scientists at the ski lodge had estimated. With the multiplex sequencing processes Church had further perfected during the last nineteen years, he now believed it would one day be possible to achieve similarly complex sequencing for a cost of a thousand dollars. The price, using Church's process, was already less than a million dollars, and it would be in the tens of thousands within a handful of years. But the work could now be done in a fraction of the time.

Church had also ventured into the new world of synthetic biology, in which scientists could sequence and then tailor simple life-forms such as bacteria to perform amazing, and sometimes useful, tasks. Bacteria could be programmed to glow like Christmas lights, to feed on waste, or even to act as biological fuel.

Along with working in multiplex sequencing and synthetic biology, Church had launched another potentially world-changing venture in 2006: the Personal Genome Project. The PGP intended to take the Human Genome Project a step further — to make it useful to the individual. Rather than sequencing a single anonymous human genome, the PGP would sequence the genomes of a large number of volunteers and make those genomes public, along with medical records and other informational resources. Eventually, having a public database of genomes, along with medical profiles, would enable medicine to be developed to address illnesses or other conditions associated with specific genomic sequences — in other words, to allow physicians to tailor treatments to individual genomes and thereby treat or cure specific individual illnesses. The project would also push the sequencing technology forward, making it even cheaper and faster.

Church himself had volunteered to be the first test subject of the PGP, allowing his own genome to be sequenced and put online for anyone to see. He'd already made his health records public. Every visit to the doctor went immediately online, along with blood tests, procedure records, and his nutritional habits. In aid of the PGP, Church was now, at the cellular level, an open book.

Over the years, Church had grown used to fielding calls about his lab's groundbreaking work, as well as informational queries from journalists working on stories about Church's many colleagues and competitors in genetics. Church had always believed in making himself available; he felt strongly that science moved faster when it took place in the open, and the more he could help educate the public, the better it would be for everyone.

So at first, he hadn't been surprised to hear from Nicholas Wade, a well-known science writer from the *New York Times,* but he was surprised that Wade hadn't called to discuss the Personal Genome Project, or any of Church's other projects. Wade had called to talk about the Woolly Mammoth.

Wade was working on an article about a team of scientists at Penn State who were about to publish a paper announcing an effort to decode the genetic material of one of the prehistoric creatures, culled from a hair sample they had retrieved from somewhere in the Arctic Circle. The scientists at Penn State believed that for around two million dollars, they could sequence the Mammoth.

Church listened carefully to the reporter, already seeing where the conversation was likely to go. Church had toyed with this idea, had even discussed something like this two years earlier with a PBS film crew, but this conversation was pushing him to consider it as a real lab project. Church was not an expert, but as a zoology major and a fan of conservation in general, he knew a bit about the Woolly Mammoth. The iconic creature had mostly died out around ten thousand years ago, succumbing to some extent to changing environmental conditions at the end of the last ice age, and hunted to extinction by prehistoric humans. Over sixteen feet high at the shoulders and weighing over twenty tons, some of them were covered in long reddish hair. Despite the most common images in popular culture, Woolly Mammoths came in a variety of hair colors, similar to the variations found in humans. And Mammoths, unlike their future elephant relatives, had evolved a special hemoglobin that could function in cells very close to the freezing point for indefinite lengths of time. Their short ears and tails resisted frostbite, and they had been supremely well adapted to the cold environments of the Arctic and the northern steppes of Siberia and North America. But even so, they had gone extinct. The last remaining handful lived on an island off the Russian coast three thousand years ago.

In the mid-1800s, numerous Mammoth carcasses had been unearthed in the glacial regions above the Arctic Circle. As the globe continued to warm and the glaciers melted more quickly in the late twentieth century and into the twenty-first, Mammoth finds accelerated. A handful of specimens had been discovered almost entirely intact.

It was a fascinating topic, but Wade wasn't calling to ask Church to weigh in on the Penn State efforts. To Wade, sequencing a frozen Woolly Mammoth wasn't enough of a story for an article for the *New York Times.* He wanted to take it further.

"Let's say they're successful," he asked. "Let's say they sequence the frozen material. Would it then be possible, using genomic engineering, to resurrect a Woolly Mammoth?"

Church smiled slightly. This was exactly the sort of thing he loved — an intellectual game, one that probably wouldn't lead anywhere concrete, but a mental thrill ride just the same.

Like many other people, Church had read Michael Crichton's novel *Jurassic Park* and seen the movie based on it. He was unlike most other people, though, in that some of his own decoding of bits of bacterial DNA had made it into the laboratory scenes in the book as so-called dinosaur DNA. And unlike most people, he knew that *Jurassic Park* was pure science fiction.

Cloning dinosaurs from genetic material harvested from a prehistoric mosquito caught in amber was impossible for many reasons. Dinosaurs had died out 65 million years ago, which meant there was no such thing as extant dinosaur DNA to be found in our modern era. No genetic material could survive even a fraction of that length of time. It would have been continuously bombarded by cosmic radiation or consumed by enzymes in the soil, which would destroy the DNA. No dinosaur fossils ever found had any genetic material. There was nothing at all to sequence. And no dinosaur fossil that would be found could contain any. An insect trapped in amber for millions of years became, at a cellular level, simply amber. It might look like a prehistoric insect, but it no longer contained any DNA.

George considered it quite unlikely that the DNA of a 65-million-year-old dinosaur would be in any condition to sequence, let alone be cloned in a laboratory. That would require adequately intact cell nuclei.

But the Woolly Mammoth was different. Woolly Mammoths were now being pulled from the Arctic ice in remarkably pristine condition, essentially flash frozen at the time of their deaths. And unlike dinosaurs, some of these Mammoths might be merely a few thousand years old.

Still, despite the near-perfect appearances

of some specimens of frozen Mammoths, attempts at growing live cells from the long-dead beasts had so far ended in failure. The DNA within the frozen cells had deteriorated over the centuries beneath the ice. In spite of science fiction writers' imaginations, scientists were not likely to be regrowing extinct creatures in a lab any time soon.

But Church wondered, what if you didn't need to *regrow* a Mammoth from a deteriorating, frozen sample? What if, instead, you approached de-extinction the same way his lab was approaching his other genetic engineering projects — with rapidly sequenced genomes and synthetic modifications to cure disease or create new bacteria? What if you could take the code for what made a Woolly Mammoth a Woolly Mammoth and implant it into one of the Mammoth's modern relatives?

Church looked down at the zoology books he'd spread open across his desk. A half dozen pictures of elephants stared back at him from their jungle and savannah habitats in Africa and Asia. On the surface, they seemed far removed from their red, furry, cold-weather giant ancestors, who had once roamed the Siberian tundra. But were they really so far apart?

Church hadn't meant to give Nicholas

Wade's question a definitive answer. He certainly hadn't intended to make any sort of announcement. He usually tried not to stick his neck out, but as an interdisciplinary scientist with a broad set of interests, Church was usually the one who got asked the crazy questions. And often, despite his best efforts, he gave the crazy answers. He always tried to be careful with journalists and to frame his answers with enough caveats to cover himself. Good journalists weren't trying to be provocative, they were simply asking what was possible, what couldn't be ruled out.

But in answering Wade's question about whether it could be possible to use genomic engineering on a sequenced Woolly Mammoth genome, Church had replied, "It's certainly possible."

And right then, he'd known he'd just given a headline to the *New York Times*.

Now, hours later, he was looking at pictures of elephants, his mind deep into what was still a theoretical game. His wife and daughter, having finished dinner, were likely heading to bed. Well, actually, Ting, being an insomniac, would surely be awake and ready to talk . . . about anything. Maybe their conversation would lead to a paper in a scientific journal, maybe it would become a thought exercise to get the hearts of the postdocs on the other side of the office door thumping. But Church was already spooling ahead.

With the sequence to the Woolly Mammoth genome, Church believed he could synthesize and implant the proper DNA code into an elephant embryo, and essentially allow a modern elephant to give birth to its own ancient ancestor.

Thirty years after *Jurassic Park,* you still couldn't cultivate dinosaur DNA from amber, but you could, if you somehow had access to a dinosaur's genome, create that same sequence of chemicals from scratch. You couldn't bring a Woolly Mammoth back to life, but you could essentially *create* one. All you needed was that genetic code and a proper flesh-and-blood incubator.

The first step was to collect the correct information. A sample of DNA didn't have to be perfect, but it had to be good enough so that you could extract the important components of a Mammoth's genetic code. To synthesize an extinct animal, you needed the proper recipe.

And that was something Church wasn't going to find in a high-tech Boston lab.

Chapter Ten

Early Spring 2009

KOTELNY ISLAND, SIX HUNDRED MILES NORTH OF THE ARCTIC CIRCLE.

Timur Khan climbed off his Russian-built snowmobile and sank almost to his knees into a snowbank. His weathered features were shaded from the high midday sun by a thick hood made of cured yak leather. His twenty-year-old Tokarov rifle was slung high over his right shoulder, the aging wood of the hilt resting close to the ammo belt around his waist. In truth, the rifle was more for show than for any real sense of protection. Though Kotelny was one of the largest islands in the world based on land mass, the ice- and snow-covered outcropping situated deep in the Arctic Ocean was mostly barren wasteland. Aside from personnel at a Russian naval base that had only recently been reactivated, and at a scientific station that was deserted most of the time, the only people Timur might run into during his hunting expedition would be Yakuts like him. His tribe was small, and he'd certainly recognize any other hunters on sight. Most likely, he'd even be related to them.

But he didn't carry a gun for encounters with other hunters, Russian soldiers, or the odd scientist. As he ran a gloved hand over his ammo belt, Timur worried about the polar bears. Especially this time of year, when the snow and ice started to build again across the region, polar bears were particularly active. And even a new rifle, one that wasn't twenty years old, or weathered by a lifetime of traversing ice floes and digging through snowbanks, wasn't going to be much use against a polar bear.

But the Yakuts were known for their courage in the face of danger. In his midfifties, Timur was not only experienced in the field, he came from a hunting lineage dating to the mid-eighteenth century, when the first Yakuts traveled north in search of better quarry.

Nearly three hundred years later, the journey to Kotelny Island was still almost as treacherous as venturing onto its polar-bear-infested terrain. First, you had to cross the forty-mile bridge of ice that connected the island to the mainland: by foot, snowmobile, sometimes by armored ATV if you could bribe your way aboard one of the Russian military excursions, or by hydrofoil if you could time your hunt with one of the rare scientific junkets to the Arctic station. Once you were on the island, you spent your time picking your way over dangerous ice floes that could crack or melt beneath your feet, past glacial boulders that came loose without warning. And at any time, you might encounter polar bears, who had been growing more aggressive every year, as the warming climate continued to shrink their natural habitats and limit their normal food supply.

Timur himself knew of three hunters who had died in the past six months. Two of them were buried within a hundred yards of where he was now standing. The third hunter's body had never been found, but Timur was certain his mutilated remains were hidden somewhere beneath the snow, ready to be unearthed with the next few digits' rise on the Arctic thermometer.

Polar bears or no, the one thing that Timur could count on was that no matter how treacherous the island, the hunters would continue to come. Before his tribe, the Yakuts of the Northern Sakha Republic, had hunted Kotelny, their ancestors had been nomadic hunters. In the Pleistocene Era, their quarry had been the great herbivores that had lived throughout the region, the caribou, bison, and reindeer that had provided sustenance. But now, the Yakuts' quarry was quite different, and infinitely more valuable.

Timur pulled his saddle pack off the snowmobile and slung it next to the old rifle. Within the pack were the tools of his trade: steel military shovels, ice picks, sealable sample containers, and a hand-drawn charcoal map. The map had been sketched by one of his cousins who had returned from an expedition to this very spot not three weeks earlier.

Timur could still picture the moment when his cousin had returned to their village after that expedition, his prize — a large tusk — strapped across his back. Their entire family had gathered around as he'd carefully unwrapped the tusk. It was only partially intact, most of its surface covered in chips and cracks, but it had to be at least thirty pounds, and nearly three feet long from its tip to where it had broken off from what remained of the carcass.

The tusk was a thing of beauty. Though browned by age and ice, it held glimpses of white beneath. It was pure ivory — but not from an elephant. It was part of a Woolly Mammoth tusk, unearthed near where Timur was standing, dug from the ancient permafrost by his cousin.

It was an incredible treasure, but, in the scheme of things, not at all rare. Timur's tribe of Yakuts had built their entire livelihood around finds like that. Though Timur's cousin's specimen was incomplete, a single Mammoth tusk could measure as much as ten feet long, and at market could bring in more than two hundred thousand dollars.

In pieces, Mammoth ivory now went for more than five hundred dollars a pound. And the better preserved the material, the more it could be worth.

More than sixty tons of Mammoth ivory were sold every year. Most of the demand came from Asia, and as much as 90 percent of Mammoth tusks ended up in China, where it was turned into jewelry and ground into various medicines. Unlike the trade in elephant ivory, the Mammoth ivory business was completely legal. More than that, the sale of Mammoth ivory was encouraged as a hedge against the murder and harvesting of endangered African and Asian elephants. Mammoths weren't an endangered species; they were long extinct. Which meant that every ounce of Mammoth ivory sold to a collector or jewelry designer or doctor in Beijing, Hong Kong, or Shanghai meant another elephant hadn't been killed for profit.

Conservationists around the world had applauded the growth in the Mammoth ivory market. But the conservationists, who faced considerable danger fighting the elephant trade, did not have to risk their lives in search of those valuable Mammoth tusks. The Yakuts had a near monopoly on the Mammoth-hunting business, due to their proficiency as nomadic hunters and their proximity to the Arctic ice floes where the Mammoths had once lived, and died, in massive numbers. Scientists estimated that more than 100 million Mammoths were buried under that ice, many of them over fifteen feet tall, with tusks that almost reached the ground.

Every day, more fossilized Mammoths were being discovered. And where one Mammoth lay, a dozen more were sure to be nearby. It was just a matter of following his cousin's map, avoiding the polar bears, and scouring the ice with his tools. If he were lucky, he would find a piece of tusk nearly as big as his cousin's prize.

Along with a tusk, Timur would also take a sample of the Mammoth's carcass, which he would carefully seal in one of the plastic specimen containers he carried with him. The containers had been given to him by a courier who worked for a laboratory in the United States. More and more often, Timur and his Yakut brethren were carrying back samples of Mammoths, along with their ivory, to be sold to the laboratories in cities such as Chicago, Philadelphia, and Boston.

Timur didn't know why the scientists were so eager for the Mammoth meat. After ten thousand years in the ice, it certainly wasn't edible or usable for leather. But Timur didn't really care why the scientists wanted it; he cared only that they were willing to pay for it.

Besides, the requests for samples from the American scientists were less strange than the offers now circulating through the Yakut villages from representatives of a group of South Korean scientists who were also interested in the Woolly Mammoth. Unlike the American scientists, the South Koreans weren't looking for Mammoth pieces sealed in specimen containers; they were hiring Yakuts themselves, transporting them to a massive excavation project, employing them for their expertise in the region and for their knowledge of the ancient beasts. The excavation project was located deep inland, up the side of Muus Khaya, the highest peak in the Suntar-Khayata range. Timur didn't know why the South Koreans wanted to dig inland, and into a mountain; Mammoths were much easier to find in the north, buried in the permafrost. But again, it wasn't a question he needed to ponder.

As he started forward through the snow, he brought his mind back to polar bears and his charcoal map. With danger around him, and ivory ahead of him, he had more important things to worry about than the whims of American and South Korean scientists.

Chapter Eleven

Spring 2009

THIRTY MINUTES NORTH OF SAN FRANCISCO.

A stretch of salt marsh runs along a serpent's curve of the Petaluma River, a secluded, fifty-acre sanctuary embraced on all sides by a living carpet of knee-high pickleweed.

George Church strolled along a dirt path that led up from where the car had first deposited his family after the short trip from the runways of San Francisco International Airport, SFO. He was momentarily by himself; Ting and Marie had fanned out in opposite directions as soon as they'd arrived, wanting to check out the natural streams and nearby little bodies of water that spotted the gentle slope leading up toward the two-story guest house. But Church was more interested in the pair of figures sitting in deck chairs on the unique building's front porch.

One of them, Stewart Brand, rose from his chair with a warm smile on his triangular face. Animated and angular, like an amiable praying mantis, long and trim, even at seventy-three, he had an overflowing level of kinetic energy. He was wearing a gray safari shirt covered in pockets and had a hunting knife strapped to his waist. Ryan Phelan, Brand's wife, still seated, her blond hair pulled up in a ponytail, exuded the intelligence and confidence of a serial entrepreneur; she'd sold at least two successful biotech start-ups in the past decade, and had her fingers in a couple more.

Church did his best not to trip on the high steps leading up to the porch. Over the course of his career, he'd been blessed to meet many brilliant people, but true innovators like Brand and Phelan were few and far between. Their meeting had come about by way of a butterfly effect starting with that phone call from the journalist at the *New York Times,* asking questions about Woolly Mammoths.

"It's different, isn't it?" Brand said, as he shook Church's hand and then gestured at the house behind him. "Took quite an effort to get it set up. There's about two thousand books in the library downstairs, and we added a whole second story for a guest bedroom. But you're not going to find better bird-watching, I can promise you that."

The compact little house, set about twenty yards away from the main farmhouse where Brand and Phelan spent their summers, was odd and beautiful. A former schoolhouse, seven hundred square feet and more than a hundred years old, it had been moved to its perch on a hill high that provided staggering views of the surrounding marsh, then outfitted with glistening, eight-foot-tall picture windows.

"Specially designed glass," Brand said. "Layered in an experimental UV coating visible only to birds."

"If you were a bird," Phelan added, "you would see the windows are painted in quite an exquisite pattern. It's called Ornilux, and it's modeled after the natural architecture of a spider's web. Spiders design their webs to keep birds from crashing into them. Turns out, the concept works for windows, too. The birds can see the patterns on the glass, and it saves their lives."

"Science modeled after nature," Brand said, gesturing around him. "Spiders teaching people how to protect birds. Pretty heady stuff."

Church took a seat next to the pair on the porch, beneath one of the bird-friendly picture windows. Brand and Phelan's primary home was a century-old converted tugboat named *Mirene,* docked in Richardson Bay, Sausalito. With its black hull and red windows, it seemed the perfect base of operations for a pair of revolutionaries.

Church could think of no better description of Brand and Phelan. In the late sixties, as the creator and publisher of the *Whole Earth Catalogue,* Brand had become a guide for those looking to live in harmony with the environment. He had blazed a countercultural trail that had inspired Church's entire generation. One of Ken Kesey's original Merry Pranksters, who were written about by Tom Wolfe in his book *The Electric Kool-Aid Acid Test,* Brand had been a proponent of the use of LSD — the "turn on, tune in, drop out" movement.

In fact, the *Whole Earth Catalogue* had actually grown out of an acid trip. After a particularly lucid experience in 1968, Brand had decided to sell buttons inquiring, "Why haven't we seen a picture of the whole Earth yet?" At the time, Russia and the United States had gone into orbit multiple times and NASA very shortly would send a man to the moon. But nobody outside NASA had seen a photo of the entire Earth, and Brand believed that just seeing the planet, from beyond, could have a powerful, unifying effect on the way people chose to live.

Unlike his fellow pranksters, Brand realized that technology wasn't the enemy; in fact, science could be a transformative force.

His simple query had led to a public campaign to force NASA to release what soon became the most famous photo of the Earth from space. Putting the photo on the cover of the first issue of his *Whole Earth Catalogue,* Brand had inspired millions with what was essentially a do-it-yourself guide, with essays on topics from physical fitness and farming to building rudimentary computers.

Over the years, Brand had only grown more convinced that technology — computers, biotech, even nuclear power — would aid his environmental revolution. In many ways, he was the ideological father to Silicon Valley, seeing technology as the proper lever to disrupt the status quo. In fact, in 1984, right about the same time Church was helping launch the Human Genome Project, Brand had launched the annual Hackers Conference in Marin County, California, where he had declared, "Information wants to be free." Years later, Church could have used the same phrase to describe his Personal Genome Project.

"The intersection of nature and science. That's where we learn to properly coexist. If we're going to last, this is how it begins. A spider's web."

The *Whole Earth Catalogue* had been the launch of a revolution, and no less a tech luminary than Steve Jobs had once likened it to "Google in print." Brand's current project, the Long Now Foundation, was no less ambitious. A philanthropic think tank, which Brand had created with computer scientist Danny Hillis, Long Now's goal was to look forward over the next ten thousand years. Instead of focusing on problems in the present, the idea was to see the world on a much longer time frame. One of the Long Now's initiatives was to create a permanent database of the world's seven thousand known languages. Another was "The Clock of The Long Now" — a three-hundred-foot-tall timepiece built in a mountain on Amazon founder Jeff Bezos's Texas property, to keep exact track of the next ten thousand years. Conservationism taken to an extreme: spiders and birds and people, living together in the shadow of a clock that would tick on for millennia. Brand was currently focusing most of his time on Revive

& Restore, a Long Now project with a mission to use genetic engineering to protect endangered species and to bring back extinct species.

"There are still some kinks to figure out," Phelan said. "The windows aren't one hundred percent bird-strike-proof, since at night owls cannot see the intricate spider-web pattern, but thankfully they've learned to stay away from them."

Church had actually met Brand through Phelan, with whom he had a more basic shared area of interest. Ten years after her 1995 start-up Direct Medical Knowledge had become the backbone of what was now known as WebMD, Phelan had founded a company called DNA Direct, offering genetic testing to customers via the Internet. By screening for preconditions for more than a half dozen diseases, DNA Direct had been aimed at the same sort of personalized medicine that Church foresaw for his Personal Genome Project. Phelan had sought Church out for the advisory board of her company. He had happily accepted and asked her to be on the board of his Personal Genome Project.

But they didn't reconnect for another few years, a short time before the trip to Petaluma, because of an odd little email Brand had sent Church, which had landed in his inbox just a few days after his phone call with Nicholas Wade.

Church hadn't been the only recipient of the email. Brand had also copied E. O. Wilson, the esteemed Harvard biologist and naturalist, who coined the term *biophilia* to describe the innate love for nature and other life that all living creatures share. Church had quickly written Brand back.

"Just a little over a hundred years ago," Brand now said, "in most places in the country, out East where you're from, you wouldn't have needed a sanctuary or a guest house on a hill. You wouldn't have needed picture windows, binoculars, or a telescope. You would just look up and see the ribbons of tens of thousands of birds, twisting above the trees. Can you imagine it? So many, they would block out the sun."

It hadn't been a Woolly Mammoth that had first inspired Brand to seek out Church and E. O. Wilson at Harvard to talk genetics. It was a bird. The passenger pigeon. As a conservationist and avid outdoorsman, Brand had always been fascinated by birds. But as a long-term, big-picture thinker, this one red-breasted bird had dominated his thoughts for the past decade.

"They say when they hit a forest," Church

commented, “it was like watching a raging fire. And there were five billion in North America. One of the most successful, populous species in history.”

Until they weren’t.

Church, Stewart, and Ryan had gone through the apocalyptic story of the passenger pigeon numerous times since that email exchange. For a hundred thousand years, at least, the passenger pigeon had been the most abundant bird on the planet, reaching a population in the billions by the early nineteenth century.

“Then they met us,” Phelan said.

No amount of spiderweb-patterned glass could have saved the passenger pigeon: A mass migration of Europeans into the North American wilderness, combined with the rise of the commercialized use of pigeon meat, led to organized shoots. By 1900, the very last wild passenger pigeon was killed. A few years later, the species was officially declared extinct.

The doomed bird was the prime model of what happened when humanity refused to coexist with its environment. Although extinction can be a natural process — and scientists estimate that more than five billion species have gone extinct in Earth’s history — humans have rapidly accelerated the process. Earth has lost half its wildlife in just the past forty years, and some scientists estimate that more than a thousand species disappear every year as a direct result of human activity. In just the past few decades, the world has lost multiple species of dolphin, the western black rhinoceros, the Caribbean monk seal, and almost two hundred species of birds, and 432 species are now at the highest risk of extinction if significant action isn't taken. For conservationists and long-term thinkers such as Brand and Phelan, extinctions represent a devastating threat to the planet. As a technologist, Brand had begun to wonder, was there a way to use the passenger pigeon as a model for reversing the dangerous trend?

Church, already on overdrive contemplating the de-extinction of the Mammoth, had responded to the email with optimism and with detailed thoughts about how they might bring back the extinct bird. It was more than Brand and Phelan had expected.

Cloning a passenger pigeon wasn't likely to be easy, because birds grow in eggs, a process that was difficult to re-create in a lab. And nobody had a frozen passenger pigeon lying around with intact genetic material. But Church believed there was enough fragmented DNA available to sequence the passenger pigeon's genome. You could then implant that genome into a modern relative, perhaps the band-tailed pigeon, a forest-dwelling relative of the ubiquitous rock pigeon that lives in cites around the world, and that pigeon would give birth to its extinct cousin.

Brand hadn't been aware that Church's lab already had the ability to change multiple genetic traits at once. Nor had he realized how quickly genetic engineering was progressing. Most of all, he was taken by Church's forward-thinking approach to science. Church believed that since genetics was moving forward at such an accelerated rate, you could start planning for innovations even before you had the capability to perform them.

Inspired, Brand and Phelan had traveled to Boston, arranging a face-to-face meeting with Church at a café near his lab. As Brand remembers it, Church walked into the café and introduced himself with a simple statement: "I'm George. I read and write DNA."

There he was, this incredibly tall man with an immense beard and wild hair, describing himself in five words. In that moment, Phelan and Brand became instant fans. Clearly, science was rapidly moving from passive observation to active creation, and this was the man who was helping make that happen.

Church had been similarly impressed. Brand had just published the latest extension of his philosophies, *Whole Earth Discipline,* and he had laid out a controversial form of conservationism that spoke directly to Church: Cities were good. Nuclear energy was good. Geo-engineering was good. It was exactly the sort of environmentalism in which de-extinction made philosophical sense.

Before long, Church shifted their conversation away from the passenger pigeon to the Woolly Mammoth. Both were keystone species that had once been plentiful, and both had been hunted to extinction. But the Woolly Mammoth, to George, was more compelling. Maybe it was the state fairs and circuses he had attended in Tampa as a kid, where he'd marveled at the enormous elephants, so gentle and intelligent. Or maybe it was just a form of speciesism. It was hard not to see the contrast — a giant, prehistoric beast, moving powerfully across the tundra, versus a swarm of red-breasted pigeons descending on crops or forests.

Ethically, both species deserved a second chance. But to Church, there needed to be more than an ethical reason to embark on such a complex project. He knew that for Brand and Phelan, it was much simpler — they were sitting on that porch looking out above the salt marsh, imagining that avian ribbon of birds threading in and out of the clouds. Church looked out and saw the living, breathing mud, the metronomic churn of the Petaluma River, the methodical progress of his daughter and wife as they picked their way through the marsh.

Church wasn't a conservationist or a philosopher. He was a chemist/geneticist, and if he was going to take a shot at a miracle, he needed a real motivation, something that would inspire a team of postdocs to take time away from whatever had brought them to his lab in Boston in the first place.

He believed it was scientifically possible to bring back a Woolly Mammoth. But why would you want to?

Why would you need to?

Chapter Twelve

April 23, 2011
KOLYMA REGION, EIGHTY KILOMETERS WEST OF CHERSKY.

It happened so fast, Nikita Zimov never had time to react.

The heavy front tire of the two-and-a-half-ton GAZ Sadko 33081 cargo truck Zimov was driving hit a patch of black ice, its giant treads spitting up plumes of gravel and frost as the rubber spun frantically, trying to find purchase. The truck's 117-horsepower diesel howled like a speared animal; then the nose of the truck twisted hard to the right, and the enormous chassis suddenly reared up, perpendicular to the road. A second later, all four tires were back down on the ice and the metal beast was spinning toward a massive snowbank, not skidding, gliding, toward what seemed like certain disaster.

Inside the truck's front cabin, Nikita's eyes went wide as he fought with the oversized steering wheel. His face was so close to the iced-over front windshield that he could see his breath on the interior of the glass. His foot was on the brake but he knew the damn thing was useless, now. Twelve thousand kilometers into the cross-country trip, traversing the largest landmass in the world, he already knew all there was to know about hydroplaning. Hell, he'd driven over so much ice in the past twenty days he might as well have replaced the tires with skates.

The first thousand kilometers of his journey had been on actual roads — real asphalt, street signs, on and off ramps, and even a handful of traffic lights. That had been a learning experience for Nikita, since, growing up at the Chersky Science Center, he hadn't had much experience with paved streets or traffic laws. Just seeing another car on the highway had caused butterflies in his stomach, and the handful of encounters he'd had with the police during the first quarter of his trip — stops for vague, seemingly arbitrary violations, most of which had ended with a warning, as well as a few crumpled rubles changing hands — had made him grateful when he'd finally reached the Ural Mountains, even though bandits frequented the roads there.

But nobody had tried to rob him, not that he'd had much for anyone to steal. Growing up in Chersky, he'd had more than his share of run-ins with thieves and was held up at gunpoint a few times, even shot at once. His father had barely seemed disturbed when Nikita had returned to the Science Center, a tear in his winter jacket from the bullet. Sergey Zimov wasn't the sort of scientist who bumbled around test tubes in a white lab coat. Doing the kind of science he did, where he was doing it, sometimes you'd get shot at.

Now Nikita was more than his son, he was his father's partner, on a multigenerational quest.

The farther east he'd traveled, the less traffic he'd seen — and the fewer people of any kind. For the past few days, the route had been empty save for him and a handful of other truckers, carting goods to the towns and cities that pockmarked the vast wilderness surrounding Kolyma and the Siberian steppes.

Nikita wasn't sure how other truckers handled the snow-and ice-covered, packed-mud trails that counted as roads here in the east. He'd had to pull over countless times already, to dig paths with the shovel he kept on the passenger seat when the snow had gotten too high, or to heave fallen tree branches and shrubbery out of the way when things looked too dense for his heavy truck to get through. And even so, the Sadko had taken so much damage along the route, it was leaking all sorts of fluids as it went, coloring the snow with oil, radiator juice, and God knew what else.

Hell, the entire truck was basically held together by duct tape at that point. There was a hole in the oil well, which he'd plugged ten days ago with a piece of wood he'd fashioned from a fence strut. Shortly after that, he'd found that one of the pneumatic brake lines had been cut by broken glass he'd driven over, which caused a major loss of air pressure. Crawling under the truck, he'd managed to seal it closed.

Nikita was desperate to get home before the old beast collapsed entirely.

But at the moment, he was fighting wildly with the steering wheel, fearing that he was going to die so close to home. A bare eighty kilometers, and he might as well have been on the surface of the moon. He had no phone in the truck, and the CB radio had conked out four thousand kilometers ago. The speedometer and mile counter had both frozen and become useless, he had no headlights or brake lights, and so much ice was caked on the bottom of the chassis that the steering had become sluggish, even under the best of conditions.

Spinning across the ice, rapidly approaching a six-foot bank of heavy snow towering above what looked to be a deep drainage ditch, was not one of those conditions.

Nikita used both hands to pin the wheel as far into the turn as he could, then continued pumping the impotent brakes. Even as he continued to spin, he grinned ruefully. His father had warned him that twelve thousand kilometers alone on the road could drive even the strongest man a little bat-shit crazy. To be honest, most people would probably have agreed that Nikita and his father, Sergey, were already far along the spectrum from oddness to madness. Nikita thought of the Sadko as an iron-and-steel version of his dad, protecting him, but certainly not coddling him, on his wild expedition. And the journey itself was a perfect analogy to the dream his father and he shared, the project that had brought Nikita — with his girlfriend, now wife, in tow — back to the Arctic Circle five years earlier, despite his every effort to stay away.

Keeping that truck alive and the journey moving forward was akin to keeping his father's dream alive, and there was no damn way Nikita had come so far only to die here, along with their dream, in a snow-filled ditch.

With his jaw clenched so tight his teeth hurt, he focused on a point in the snowbank, coaxing the steering wheel another few inches into the turn. If he hit the bank just right, maybe he would keep the truck from flipping over. Even without the speedometer he knew the beast had picked up speed once the tires had touched the ice. There would be no slowing until the crash. But if he was lucky, the Sadko would stay upright. That was really all that mattered.

Even as he spun wildly across the ice, Nikita could feel the weight of the cargo shift in the truck's bed behind him, separated from the back of his head by thick metal. The cargo was so heavy that the front tires were riding slightly higher than the rear, certainly affecting the four-wheel-drive suspension. There had been no way to secure the cargo once he'd loaded the flatbed, around the halfway point of his trip. For much of the time — and mostly at night — the freight stayed toward the back of the bed, putting much of the weight on the Sadko's tail. But when the ride got agitated — well, it was like driving a truck full of wild animals.

Nikita grinned at his private joke. Even as he headed toward the ditch, he recognized the absurdity of the situation. Many young Russian men of his age went on cross-country road trips, to see the great nation, its incomparable landscapes, big cities, and the modern world that had replaced the staid Soviet era. The gray authoritarian state had turned into a vibrant, culture-filled country.

But Nikita hadn't set off on his long trip to see the world. He had set off to *save* the world.

He had begun his adventure with a sense of happy anticipation. Flying from St. Petersburg to Nizhny Novgorod via Moscow, he'd gone from one Soviet-era airport to another, marveling at the terminal buildings of stone, matching spiral staircases, and balconies that had once been emblazoned with enormous posters of Lenin and Stalin. But in Novgorod — a large city on the banks of the Volga River that was a tourist destination — most of the terminals were stark and modern, with flashes of Western-style advertising, touting products like soft drinks and cell phones. Shortly after landing in Novgorod, he'd picked up the truck, which his father had reserved and paid for.

When Nikita had first seen the Sadko,

parked in an alley behind a dealership trading in old army castoffs and heavy construction material, he'd nearly wet his pants: five thousand intimidating pounds of military-grade diesel charm, with tires that came up to his waist and faded, dull green paint. A huge flatbed made up most of the truck's length and was covered in a tight canvas tarp, which stretched at least six feet high over several curved iron bars.

When Nikita first climbed into the truck, he'd felt twice as tall as usual, and strong as a musk ox. It took him most of the first day of driving the beast to learn how to control it, but once he was out on the roads, he quickly got used to the strength of the engine and the pull of the manual steering.

Luckily, by the time his cargo joined him, he'd mastered the truck and its idiosyncrasies and steeled himself for the possibility of encountering bandits in the desolate mountain roads of the Urals.

Nikita closed his eyes and yanked the wheel the last few inches, just as the first tire slipped off the edge of the road and into the ditch. He felt a sudden, frozen moment of weightlessness, and then the truck was pitching to the left, the right tires lifting off the ground. Nikita started to scream, but was cut short by a hard thud as the Sadko slammed back down and skidded into the bank. The truck came to a hard stop, two tires sunk in the ditch, the other two spinning in the air.

Nikita opened his eyes. The front cabin was tilted forty degrees, and his body was jammed against the driver's-side door, which was flush with the snowbank. His hands ached where they had gripped the steering wheel, his knee was going to have a nasty bruise from slamming into the door handle, but nothing felt broken. Hell, better than that, he was alive.

And then he heard the banging from behind him in the flatbed, followed by an eerie, high-pitched scream.

His cargo.

Nikita hoisted himself across the front passenger's seat and kicked open the door. Then he took hold of the door frame to pull himself out of the cabin. He balanced on the rim and dropped down onto the ice. His boots hit with a satisfying crunch as he stuck his landing and stood up.

Then the cold and the wind hit him, and he shivered violently, from the adrenaline as well as the temperature. It had to be below freezing outside, and he wasn't wearing a coat. But he didn't care; his mind was entirely on his cargo. Besides, he couldn't have felt more Russian than at that moment, standing on the edge of an icy Siberian road, his truck in a ditch, wind whipping through his hair. He only wished he had a bottle of vodka to finish the picture. There were very few moments that a bottle of vodka wouldn't make better.

Nikita moved quickly along the side of the tilted truck, until he was parallel with the cargo bed. The canvas tarp was torn in many places, but was mostly held intact by the iron bars. Only a small section near the front cabin seemed to be tearing loose from one of the bars.

Nikita braced one foot on the tire guard and pulled himself up the side of the truck. He reached both hands out toward where the canvas had come loose, his fingers getting closer — when suddenly, the sharp point of a huge antler tore through the tarp just inches from his chest. Nikita twisted out of the way just in time, but as the antler pulled back into the flatbed, it pulled more of the canvas with it, opening a large hole in the covering.

Now Nikita had a full view of the interior and the six enormous elk huddled together near the back of the bed. The bull elk — the dominant male — was standing in front of the others, balancing its three-hundred-pound frame against the tilted truck bed on long, almost spindly legs, looking up at him from beneath a huge rack, his eyes wild. Condensation clouds curled up from his arched nostrils, and saliva dripped from his mouth.

The attempted goring hadn't been malicious. The bull was simply protecting his herd and posturing to keep his position.

With a loud grunt, the bull elk leaped forward again, its horns coming straight at Nikita, who lurched backward, putting himself behind one of the curved iron bars that kept the canvas in place. The antlers hit the bar and the animal fell back again, shaking the entire truck as he landed.

Okay, maybe the bull *was* trying to kill Nikita, but it had pretty good cause. For twelve days, Nikita had been carting the six elk across the Russian continent, feeding them, cleaning out their waste, keeping them healthy as best he could. Driving seventeen, eighteen hours a day, stopping only for brief periods of sleep and to buy food, fix the truck, acquire diesel — and now he'd nearly killed them all on a patch of ice.

The elk didn't have the capacity to understand, but their trip and their relocation were integral to Nikita and Sergey's mission.

Nikita stretched forward again, pinning the canvas as tightly as he could against the iron bar. Again, the bull elk came at him, those sharp antlers slamming into the iron, sending up icy sparks. Nikita laughed out loud, from adrenaline and joy. The bull was perfect, indomitable, strong like Nikita and Sergey. The Arctic Circle — the permafrost steppes where he and his herd were headed — was not a place for the weak. Only the strong could survive there; only the strong could help rebuild what once was, to help turn back the clock . . .

Ten thousand years.

The elk bugled — something between a high-pitched scream and a howl — as it pounded up the truck bed again. Nikita howled back at it, holding the canvas with all his strength. No matter how long it took for another trucker to drive by or for his father to come looking for him — no matter how long it took for help, in some form, to arrive — he'd hold that canvas and keep the six elk safe and secure. Nikita would deliver them to the refuge he and his father were creating above the Arctic Circle, and together, these elk, Nikita, and his father would continue working toward saving the world.

Hell, even vodka couldn't have made this moment more exhilarating.

Chapter Thirteen

October 24, 2012

HUBBARD HALL, THE CORNER OF SIXTEENTH AND M STREETS, WASHINGTON, D.C.

On the second floor of one of the most significant buildings in the history of science, an elegantly appointed conference room and meeting hall were tucked between a pair of libraries. Their huge arched windows looked out past austere Doric stone pillars to the busy streets of the nation's capital below. Parquet hardwood floors gleamed, polished and smooth, beneath a fifteen-foot ceiling of ornate plaster moldings and marble. Gathered together in groups of three, four, and five were thirty-six of the smartest men and women in biology and conservation, many meeting each other for the first time, to talk shop.

Church could think of no more fitting setting for inspiration. In 1909, Hubbard Hall had housed the first headquarters of the National Geographic Society, and it was still the philosophical center of the organization that had been at the forefront of scientific exploration for well over a hundred years. That National Geographic had invited Brand and Phelan to hold their conference in such a legendary place signified that they were conducting groundbreaking science. The workshops and presentations Church had taken part in so far had stoked his enthusiasm for resurrecting extinct species, and he couldn't help thinking that the energy in the place was reminiscent of those days in Alta, where the Human Genome Project was hatched.

When Phelan and Brand had first set up the conference, they had billed it as a private workshop involving the top scientists working on de-extinction and related projects. The goal was to continue the conversation that they had started in Petaluma, melding molecular biology with conservation biology to see where such a marriage could lead.

Church had gone into the conference with an open mind. He didn't necessarily expect to find a reason for turning his imaginings about resurrecting Woolly Mammoths into something concrete and real, but if a reason for de-extinction existed, he knew he'd find it here, by mixing with some of the world's top biologists and geneticists. Church's daughter, Marie, now a professional photographer, was present for the event and had generated the official photograph that went out to the press afterward.

For two days, he had attended a cavalcade of fascinating presentations, beginning with Alberto Fernandez-Arias, a veterinarian who was also the head of the Hunting, Fishing, and Wetland Department in Aragon, Spain. Fernandez-Arias spoke about the only true successful cloning of an extinct species, the Pyrenean ibex, to date. The story was both stunning and frustrating.

In 2003, a Spanish team had managed to clone a recently extinct Pyrenean ibex — a sort of mountain goat — from a tissue sample that had been frozen in liquid nitrogen since the last of its species had died in 2000. Placing the genetic material into the eggs of domesticated goats, they'd managed to get a single pregnant mother to term, and a living baby Pyrenean ibex had been born. Unfortunately, the animal had lived for only ten minutes, before suffocating because of a deformed lung. But for those ten minutes, an extinct species had indeed lived again.

Of course, at the time, the ibex had been extinct for only a little over three years, and its genetic material had been carefully preserved in a laboratory. This was no ten-thousand-year-old Mammoth carcass dragged out of Arctic ice. But it was an impressive accomplishment, nonetheless. To Church, it was on par with the much more famous cloning of Dolly the sheep — the first mammal ever cloned. Although Dolly had lived seven years — and had even given birth to seven lambs before dying of lung cancer at the age of seven — Dolly had not been a member of an extinct species, nor had she been cloned from tissue preserved in a lab.

Church's own presentation was relatively down-to-earth, although it was a surprise to many of the attendees. Onstage, he reiterated much of the conversation he'd had with Brand and Phelan, telling the gathered scientists how far his lab had come in rapid genetic sequencing and cheaper, faster genetic engineering. Within the next five years, he explained, they would be able to make over a dozen edits to a three-billion-base-pair genome for only thirty thousand dollars. His lab's DNA editing breakthrough — called MAGE, for Multiplex Automated Genome Engineering — would eventually allow them to replace genes in a living species with similar genes from an extinct one, and this would one day make it possible to implant Woolly Mammoth genes into its closest relative, the elephant.

The scientists were excited to learn of Church's advance. These were the tools that would make their own work possible. It made the presentations that followed, regarding the de-extinction of the passenger pigeon, the Tasmanian wolf, the European aurochs, and others, seem much less like science fiction, much more like future fact.

But Church was most excited by the presentation by Sergey Zimov, who had come from the edge of the world, Siberia, at Brand's invitation.

Church hadn't been familiar with Zimov's work and had needed to look him up online. But the minute Zimov took the stage, Church realized that his message was going to be something different, unexpected.

Zimov's appearance was itself a spectacle. Long gray beard trimmed to a sharp point, noble, broad features lined by weather, hardship, and age, he looked like a figure who might have been painted on czarist-era Russian nesting dolls. Zimov wasn't fluent in English, so his words were filtered through a competent translator. He made it clear from the start that, though he might look like a reclusive scientist, he wasn't a loner. His son, Nikita, was a partner in their multigenerational mission.

And Sergey and Nikita's work was spectacular. In a world well north of where most people could survive, Zimov conducted his research on an expanse of steppes surrounding the Chersky Science Center, where howling winds often reached sixty miles per hour, and temperatures dropped below minus ninety degrees. There, over millennia, species after species had gone extinct, and even the ground itself had gone from plains teeming with thick, sustainable grass to jagged permafrost strangled by moss, weeds, and lichen.

That permafrost stretched for tens of thousands of miles in either direction, crossing multiple continents, circling the entire globe. *A landmass covering as much as 20 percent of the Earth's surface.* It was solid, seemingly impervious, perhaps ten, eleven feet thick. And this permafrost, which circled the world like an icy crown, held a devastating secret: It wasn't invulnerable at all — it was a ticking time bomb. Locked in a fragile balance with a worldwide ecosystem that was shredding at the seams, it could release enough carbon to tip the planet into an irreversible global warming.

Hours after Zimov's talk, Church was mobilizing to deal with the dark future the Russian had described.

Global warming, climate change, whatever words humanity chose to define the slow drip of the carbon-powered faucet that was drenching the world, had hit the Arctic especially hard. Data from space satellites and ice sensors showed that the area was warming at twice the rate of the rest of the world. The sea ice that had surrounded the polar cap since the last ice age would melt entirely during some summer season within a single generation. A rise in water level due to the melting glaciers was less of a threat than a more insidious danger, one that began just inches below where Zimov and his family spent most of their days.

The tundra wasn't just ice and rock; the permafrost that stretched around the cap of the world contained massive pockets of methane and almost three times more carbon than all the forests on Earth combined. As the Arctic warmed, and the permafrost began to melt, that carbon dioxide and methane would be released — a trickle at first, but once it gained momentum and passed a certain point, a feedback loop would engage. That carbon dioxide would billow out into the air, warming more and more of the permafrost, which in turn would release even more carbon. That trickle would become a toxic flood. Eventually, the permafrost would release more carbon than would be created by burning all the forests on Earth three times over.

The effects would be disastrous. The melting permafrost could suffocate the world.

According to Zimov, there was still a chance to hold back the feedback loop. Zimov's data, which he'd been painstakingly collecting for decades, proved that the tundra of his home, as forbidding as it seemed, could change. And the key he described wasn't some futuristic technology; quite the opposite, the key came from the distant past.

Church had circled the meeting hall twice, and still had found no sign of the Russian scientist. Perhaps he had already started his long journey back to Siberia. Zimov didn't leave his home often. In the past three decades, the man had not been out of Siberia more than a handful of days. Maybe that was the reason his experiments had taken so long to reach the outside world — why they were still unknown to the majority of scientists. Why his elegant solution to one of the most frightening consequences of climate change seemed completely new, even though it was, at its heart, a reversal of time, turning back the clock by millennia.

Twenty thousand years ago, Zimov had explained, the tundra was very different than it is now. During the last ice age, the last great global freeze which marked the tail end of the Pleistocene Era (which stretched from 2,588,000 years ago to 11,700 years ago, when our anatomically modern human ancestors emerged), the tundra wasn't a scarred bed of moss and lichen; it was a lush refuge of high grass. Megafauna — herds of giant, furry herbivores, from horses and buffalo to reindeer and Woolly Mammoths — populated the steppes in huge numbers, continually trampling and turning the topsoil above the world's largest biome as they grazed. Even as the ice age ended and the world began to warm, the herbivores naturally tilled the earth, churning the soil to expose the frozen ground beneath to the even colder air, keeping the permafrost perpetually chilled.

Zimov believed that the megafauna could have survived the changing climate. Their own grazing and foraging activity kept the topsoil perfect for grasses, while preserving the permafrost beneath. But as the glaciers receded, an even greater danger emerged for the animals of the steppes. As the air warmed, the newest mammalian inhabitants moved north: tribes of humans, with an appetite well beyond that of any other predator that had ever reached the tundra. By the end of the Pleistocene Era, a mass extinction had begun. The megafauna were hunted to extinction, and with them went the ecology of the finely balanced system. The grasslands withered and died. The moss and lichen took over. Trees sprang up haphazardly between the weeds. And over time, the permafrost began to melt.

The time bomb began to tick.

But Zimov believed — and had convinced Church, Brand, and Phelan — that there was a way to slow the ticking. Maybe even stop it for good.

Pleistocene Park.

Even the name sent pleasurable chills down Church's spine. Zimov had started the project almost thirty years ago, in 1988, beating *Jurassic Park* novelist Michael Crichton to the punch by two years. Simply put, Pleistocene Park was a conservationist's attempt at time travel. Using 160 square kilometers of Siberian tundra given to him by the Russian government, Zimov's goal was repopulating the area with modern equivalents of prehistoric animals that had adapted to Arctic conditions. As the herbivores turned and stomped the topsoil, continually exposing the permafrost to the cold air and wind, the permafrost beneath would cool, preserving the precious permafrost — in the warmer months, knocking down trees and curating the grasses, increasing reflectance of the surface (the albedo effect), and restoring the ecology of the late Pleistocene.

Working within the limitations of his Soviet-era budgets, Zimov had managed to stock his Arctic refuge with moose, Yakutian horses, Finnish reindeer, and even North American bison. Now with the help of his son he'd brought in more animals — elk, musk oxen, and special breeds of yak. Although his herds were still small, sometimes fewer than ten members, he mimicked larger herds' behavior by using tractors, pile drivers, and bulldozers. To re-create the effects of Woolly Mammoths on the land, he'd brought in a World War II tank, which he'd bought off a defunct military base and driven hundreds of miles across Siberia to his home. Punching holes in the snow, bashing away rocks and trees, churning up the moss and lichen, using the tank treads to mimic the continual stomp of Mammoth feet, he'd worked the land, year after year.

And along the way, he'd accumulated data that were staggering in their implications.

Within his 160-square-kilometer refuge, he had lowered the permafrost temperature by an average of fifteen degrees. Church believed that Zimov had sufficiently proved that the megafauna of the Pleistocene Era had lived in balance with their ecology, and that a reintroduction of similar megafauna could sustain that ecology for the foreseeable future. Zimov's "laboratory" was small, a tiny percentage of landmass compared to the whole, but if he could duplicate his Pleistocene Park on a larger scale, he could keep the permafrost from melting for decades.

The Russian had found a way to defuse the ticking time bomb.

In Hubbard Hall, Church realized that he wouldn't be able to catch up with Zimov. He would have to communicate with him in a different manner: through the tools and inventions of his craft, through the science of his unique Harvard lab.

There were only so many bison and Yakutian horses Zimov and his family could buy and transport to their refuge at the top of the world. Their budgets were limited, and their data were just that — numbers on a page.

But their moose, horses, bison, reindeer, and elk had proved that Pleistocene Park could work. Now they needed something much bigger, something much more ambitious to capture the world's attention.

As Zimov had said at the end of his presentation, "To fight the forest, instead of Mammoths we now use military tanks. Unfortunately, they don't create dung."

The rest of the scientists had laughed, while Church exchanged looks with Stewart Brand and Ryan Phelan.

The Russian scientist had just given them their reason to resurrect their species.

Chapter Fourteen

EXCERPTED FROM "THE WILD FIELD MANIFESTO" BY SERGEY ZIMOV

For hundreds of millions of years, terrestrial ecosystems were an arena of struggle between plants and herbivores. To avoid being eaten, plants protected themselves with thorns, tall heights, bitterness, acids, and sharp smells. Many developed numerous poisons: Solanaceae developed nicotine; poppy developed morphine; and willow developed aspirin. But, 20 million years ago, life on the planet changed. Grasses and quickly growing pasture herbs appeared. They did not spend resources on thorns and poisons; their main strategy was rapid growth. All of them were tasty, nutritious, and not afraid of being eaten. Giving several harvests a year, these plants fed numerous big herbivores. What wasn't eaten by "bulls" and "horses" was eaten by omnivorous "sheep" and "goats." In this way, the evolutionarily youngest ecosystems were formed — pasture ecosystems.

Like in economics, in ecology the rate of the capita turnover is important. According to the V. I. Vernadsky law, evolutional processes are directed towards increasing the turnover of biological elements. For instance, in the evolutionarily relict spruce forest, bio-cycling is slow and green leaves live for ten years. This biomass is barely edible and decomposes slowly on the soil surface. In contrast, grasses in pastures live only a few weeks on average. In the warm stomachs of herbivores, they decompose in just one day and their main ecosystem capital (nitrogen, phosphorus, and potassium) is quickly returned into the soil, and eventually into new leaves.

These rapidly growing grasses needed abundant mineral supplies, which herbivores themselves maintained. Abundant herbivores managed and extended their pasture ecosystems themselves. Moss and lichens were trampled. Goats and roes ate the young trees and shrubs seedlings. Bison and deer killed trees by eating the bark. Elephants and mammoths simply broke trees. Through fertilizing, harvesting, and trampling, herbivores managed their pastures in any climate.

Fifteen thousand years ago, pasture ecosystems were at the peak of their evolution.

They occupied most of our planet. . . .

Fourteen and a half thousand years ago, sharp climate warming took place. The Ice Age was over. Human chances for survival, especially for kids, substantially increased in the middle and high latitudes. People populated north of Eurasia and then penetrated into America. Experienced and well-armed hunters met herds of untamed animals. The further humans moved from their historical motherland, the more they engaged in their "blood-thirsty pursuit."

In northern Asia, 8 megafauna species went extinct upon human arrival, in North America, 33, and in South America almost all — 50 species in total. As hunting and technology developed and animal density on the pastures declined, the animal density in most regions became insufficient to maintain pastures. As a result, forest and tundra (shrubs, trees, moss) began to press pastures, causing forest area in the world to increase ten-fold. . . .

Of all the wars humankind has fought in the past, the war with pasture ecosystems is the longest lasting one. But today it can and must be stopped. . . .

Most of the species that once roamed in the pasture ecosystems have survived — some in the forests, some in deserts, some in the zoos, and some as domestic species. Other animals are proposed to be re-created through genetic engineering. All that is required to re-create pasture ecosystems is to reliably fence off a territory where grasses and herbs grow. The second step is to collect all animals that can live on this territory. Once there, the animals will remember how to live with each other themselves. They will divide pastures and occupy all ecological niches according to their professions. It is their job to self-regulate density; the weak will die, the strong will re-populate, the ecosystem assemblage will stabilize and, then, will be ready to reintroduce into new territories. . . .

Frozen soils of the mammoth steppe contain lots of organic carbon — three times more than all tropical forests of the planet. When these soils thaw, microbes that were previously frozen there for millennia wake up and immediately start to decompose the soil organics, producing the greenhouse gases — CO_2 and methane. If current climate change continues, then in the not too distant future the thawing soils of the mammoth steppe will be the biggest natural source of greenhouse gases on the planet. This will cause additional warming to the climate and permafrost will thaw even quicker. We cannot artificially stop this process.

However, pasture ecosystems can. . . .

Animals in pastures, looking for food, excavate and trample all snow several times each season, causing it to condense and lose its heat-insulating abilities. Therefore, the introduction of animals on pastures cools permafrost temperatures by 40 C, which can stop or substantially slow down permafrost degradation.

Forest and shrub lands are dark year-round and absorb the sun's heat well. Pastures are much lighter and, in the winter, are white if they are covered with snow. Therefore, pastures reflect more of the sun's heat and cool the climate.

It is very hard to agree to reduce industrial CO_2 emissions. Reducing permafrost emissions is much easier. All that is needed is to cross mental barriers, accept that pasture ecosystems have a right to live and to freedom, and return part of the territory that our ancestors took from them.

▪ ▪ ▪ ▪

Part Three

▪ ▪ ▪ ▪

I like to keep the median age in my lab low so we can dream together and make those dreams come true. They don't yet think things are impossible.

— GEORGE M. CHURCH

It's all too easy to dismiss the future. People confuse what's impossible today with what's impossible tomorrow.

— GEORGE M. CHURCH

Chapter Fifteen

Winter 2012

77 AVENUE LOUIS PASTEUR, NEW RESEARCH BUILDING, HARVARD MEDICAL SCHOOL.

Luhan Yang was moving fast as she navigated the crowded hallway that bisected the third floor of the New Research Building. Although everyone walked quickly at Harvard Medical School, Luhan was a bullet cutting through the stream of med students, lab technicians, and professors, determined not to be late to the open afternoon seminar on knockout genes and antimalarial mosquitoes. Even as she went — the crowd parting in front of her rocketing five-foot-three frame — she had her cell phone out in front of her, and her eyes were focused on the screen. The moving picture on the screen had nothing to do with mosquitoes. It was a video of an ultrasound, taken just a day earlier. Peering closely, Luhan could even make out the tiny, beating heart. The sight of something beautiful and precious and perfect, suddenly alive, gave her the ghost of a smile.

Luhan didn't show emotions often. She was an exceedingly efficient person, even in the way she walked. Usually, her limbs moved only the exact amount necessary, and at the moment, her jet-black hair was tied in a gravity-defying bun. She had a sharp sense of humor, and she could relax and let go in the right setting. But, day to day, she simply didn't see the point of expending energy on unnecessary emotions.

"Hey, congrats!" shouted a med student as she passed Luhan and glanced at the screen over her shoulder.

"Thank you," Luhan responded.

It wasn't until she'd gone another five feet that she realized the medical student thought the baby was hers. It wasn't. In fact, it wasn't a human baby at all.

It was a pig. That wasn't entirely accurate either, because although most of the DNA of the fetus was indeed porcine, there was a little bit that . . . wasn't. The pig fetus — named Laika after the first dog that the Russians had sent into space — had a liver that was partially human. *Or more accurately, human compatible.* And that little piglet, Luhan believed, would one day change the future of transplantation medicine.

Luhan didn't feel any sense of hubris at the ambitious thought. Two years at Harvard had pretty much inured her to outcomes that might have seemed extraordinary at any other lab. She had done groundbreaking research (with another postdoc, Prashant Mali) on genetic engineering in human cells, which had caught the attention of a pair of transplant surgeons. The surgeons had been working for decades, mostly unsuccessfully, on using pig organs for liver and kidney transplantations, and now they wanted to see if the work Church and Luhan were doing could make possible what so far had eluded them.

To the uninitiated, Luhan could come off as intense. If a fellow student couldn't understand something she was trying to explain, she didn't coddle him. She had a great capacity for empathy, and knew that an iron mind alone wasn't enough; as a woman succeeding in science, she knew it was important to be accessible, to do her best to inspire others to pursue their dreams. But sometimes, she didn't keep it a secret that ignorance and laziness frustrated her.

In that, she'd found a veritable paradise in the Church Lab. Everyone at the lab was smart, or at least had a baseline of knowledge that Luhan could work with. And she was at the top of the food chain, a favorite of Dr. Church. When a group of bored postdocs had hacked into Church's daily schedule and had written a computer program to crunch the time sheets to see with whom Dr. Church spent his time, Luhan had dominated the rankings. Week after week, the only people Church spent more time with were his own family members.

To be fair, part of that time Church had spent teaching Luhan English. When she'd first arrived from China, her knowledge of the language had come entirely from watching dubbed television shows that classmates had smuggled into the dorms at Beijing University, which meant she could handily describe the bathing suits on *Baywatch,* but very little else. She could barely communicate outside of her lab work, which hadn't been helpful when she'd first interviewed for Harvard's Ph.D. committee. And later, her graduate school qualifying exam committee had nearly turned her down, despite the fact that she had been the number-one-ranked student in high school, and one of four kids chosen to represent the entire country of China at the month-long international Biology Olympics in Australia. There was an actual statue of her standing in a park in her home town, where intellectual prowess was considered as celebrity-worthy as traits like athletic ability or physical beauty.

But Church, perhaps seeing a parallel with his own rejection at the start of his Ph.D. studies, had stepped in with an offer to teach her as she worked. His lessons had revolved around the scientific project she'd joined his lab to pursue, but he'd pushed her, and she had learned. For Luhan, it was the perfect situation; she could learn the language without wasting valuable lab time taking a language course, and she would get one-on-one time with the most brilliant mind in genetics.

The modern American hallway she was now churning through was a stark contrast to the rural mountain town deep in mainland China where she had grown up. Luckily, her father had worked for the government, which had ensured their family a certain level of prosperity. When China had shifted from a purely communist state to a country that embraced private enterprise, her father had gone from being a public servant to the director of a factory. To make sure his transition went smoothly, the first person he had fired from the factory was Luhan's mother. At the time, Luhan hadn't understood that her father had simply chosen the expedient political route. But her mother had thrived outside work, teaching tai chi and helping to make Luhan's childhood as worry free as possible.

Luhan hadn't seen herself as special and was as surprised as anyone when she'd graduated number one in her middle school class. At her parents' prodding, she then applied to the best high school in China — a boarding school in the capital of Szechuan Province. Out of eight thousand applicants, fifty from outside the capital were accepted, and Luhan made the cut.

Life in boarding school hadn't been easy. As a country girl, Luhan had been an outsider, an easy target for her more cosmopolitan classmates. She was teased for her rural accent, and she learned to tamp down her sensitivities, to cover her emotions as much as she could.

In her second year of high school, when she went to Australia for the Science Olympics, she saw it as an opportunity to move beyond the walls of boarding school — to interact with foreigners, professors, people who didn't care where she came from, only how she thought, what she could do.

After two more years of high school, she was on to Beijing University, where she finished as the top student in the department of life sciences. She'd never thought she would stay in China after university, but around that time, her mother became extremely sick, so Luhan returned home. Her mother faced a difficult decision — should she have surgery to remove part of a lung or undergo treatment with a medicine that would leave her sick and helpless for a long time? Knowing that the second option would force Luhan to stay in China to help take care of her, Luhan's mother chose surgery, telling her daughter to go out into the world and be useful. That, to Luhan's mother, was the more important thing.

Harvard, and the Church Lab, seemed the perfect destination. For her predoc work, in 2009, she'd pitched Church on a project involving chimeric dreaminase proteins to be used for flexible, safe DNA engineering. A chimera was an ancient, mythic beast that had the body of a lion, the head of a goat, and a tail that ended in a serpent's head. Today, with bioengineering, scientists can create proteins and even DNA that contain parts of separate and distinct species. Whereas the current process of using bacterial DNA to cut all the way through genes could be toxic to the cells, a dreaminase protein could change itself to match up with the damaged genetic material, which would allow safer, more precise binding.

Normally, a first-year graduate student at a gene lab wouldn't have been allowed to launch a project like that on her own, but Church had been immediately impressed by Luhan's ideas. Now, two years later, moving down the third-floor hallway toward the seminar, her phone still raised in front of her, she felt that familiar presence hovering over her left shoulder well before his long shadow winked across the ultrasound on the screen.

"He looks happy," Church commented, matching her speed. At his height, he was taking one step for every two of hers, but now that it was the two of them, people were getting out of the way much faster, which made their progress much easier.

Luhan allowed herself to smile. When she'd first started at the Church Lab, she'd been a jumble of nerves whenever they spoke. But as her English improved, and she'd learned that Church was warm and nonjudgmental, in contrast to many of the professors she'd known in Beijing, she'd grown comfortable with him.

"He is very healthy," she said. "Although we won't really know very much until he grows up, and we biopsy his liver."

Church wasn't only supervising Luhan's project, he was helping her to get a patent for this very special pig, and also to form a company around the process that created the piglet.

Church seemed to slow a step, and Luhan had to choose between cutting her own gait in half and sprinting ahead.

"Do you have a moment?" Church asked.

"The seminar . . ." she said.

"I think it can wait."

Luhan had been interested in learning about the lab's progress on the mosquito project, for which they were nearing the testing phase. The Gates Foundation, run by Bill and Melissa Gates, had already been investing in huge domes over a trio of villages in sub-Saharan Africa, where the very special mosquitoes that had been created in Church's lab would be released, to be tested in a safe environment, one from which they couldn't escape.

But Church was already leading her toward a door set a few feet away in the third-floor hallway. Luhan found herself in an empty classroom, following Church through a maze of chairs that formed a half circle facing a desk at the center. Church pulled an iPad out from under his arm and placed it on the desk so that she could see what was on the screen.

In the center was a painting of three Woolly Mammoths, walking through a pasture of grass and snow. They looked much as she remembered them from the pictures in biology classes in high school. Thick red hair, huge tusks, small, rounded ears, long, curled trunks. Then she noticed that above the Mammoths was a line of bold text, a title or headline, of sorts:

WOOLLY MAMMOTH REVIVAL.

"This is something I've been working on," Church said, "with Stewart Brand and Ryan Phelan. A new project that we hope to take live as soon as we figure out a team and the proper funding."

Luhan looked up at him.

"A website?"

Church paused.

"No, the Woolly Mammoth. We're going to try and de-extinct a Woolly Mammoth."

Luhan slowly slid her cell phone into her pocket, as Church directed her to a page of text below the picture of the Mammoths. It began with a mission statement that would have fit nicely in the prologue of a science fiction novel:

The ultimate goal of woolly mammoth revival is to produce new mammoths that are capable of repopulating the vast tracts of tundra and boreal forest in Eurasia and North America.

Luhan skimmed the rest of the text: The story of Sergey Zimov's research showing how the tundra, with its ticking time bomb of carbon and methane, could be brought back to a state of Pleistocene pastures by the introduction of large, prehistoric herbivores. How a herd of Woolly Mammoths, engineered in a Harvard lab in Boston, might save the world from the melting permafrost.

Luhan allowed herself to digest what she was reading. The lab was filled with crazy projects that were somehow very real. It was the type of place where you had to set your own limits, stay focused. By staying focused, she had completed her own initial project on dreaminase proteins in twenty months. She had gone from that work into the pig organ transplantation project. Luhan loved that work, because it was exactly what her mother had wanted — for her to be useful. Thousands of people died every year waiting for livers and kidneys; in China, the situation was even worse, because according to their local Buddhist beliefs, people needed to die intact, and the concept of organ donation was troublesome.

The ultrasound on the phone in her pocket was the definition of useful science. If she and Church succeeded with the project, they could save thousands, perhaps tens of thousands of people. Yet bringing back the Woolly Mammoth would solve an ecological problem that affected billions. It would also be a vivid demonstration of what genetic science could do for the world — the way her pigs, once impossible, were now possible.

And on a much bigger scale.

"I want you to head the Woolly Mammoth team. Be our first Revivalist."

Luhan understood what Church was asking. Church, as the head of his lab, supervised, inspired, pushed, prodded, and shepherded the postdocs who did the experiments. But the postdocs ran the projects themselves. A good professor chose the smartest people he could find, set them in a direction to accomplish a task, then got out of the way, stepping in only when they ran into a roadblock they couldn't clear alone.

"The Revivalists," Luhan repeated. "Like the Avengers. Superheroes."

Maybe it was her English, but it sounded funny to her, even though she knew that

Church was dead serious. This was something he was going to attempt, with or without her. And as crazy as it seemed, she knew she was already in.

She also knew that the first call she'd make when she got home was to her parents. Her mother wouldn't see this as science fiction at all. Once, after Luhan had first arrived at the Church Lab, she had told her mother that she was going to make a dragon by taking a snake and inserting genes for legs, wings, and feathers. Her mother had believed her, for which Luhan forgave her. After all, there was a statue of Luhan in her home town. Her parents believed she could do anything she set her scientific mind toward.

"We can do this?" Luhan said, her voice somewhere between a question and a statement.

"I think we can," Church said.

"*Should* we do this?"

People in Church's lab asked that question frequently. Luhan thought about the seminar she was missing about the mosquitoes that had been invented in their lab — mosquitoes that had been engineered to be impervious to carrying malaria, and engineered to spread throughout the African ecosystem. The mosquito team would move forward as carefully as possible, from their domed test villages into the world at large. Eventually, they would save millions of lives.

Luhan looked back at the family of Woolly Mammoths, already thinking about who she was going to ask first to join her team. Who was going to be the next Revivalist.

Then she turned back to Church.

"I think we're going to need a very big dome."

CHAPTER SIXTEEN

Winter 2012

MACDONALD-CARTIER FREEWAY, HIGHWAY 401 EAST, U.S.-CANADIAN BORDER CROSSING INTO MICHIGAN.

A little after noon, Bobby Dhadwar stared out through the windshield of his Honda Civic hatchback and counted the cars ahead of him. Four, five, maybe seven, and then the rest of the queue disappeared into the swirls of snow blowing back and forth across the highway. Beyond the line of cars, he could just make out the border patrol booths running perpendicular to the stretch of pavement, divided by bright red electronic gates that separated his old home from his new. Each time one of those gates rose, each time one of the cars in the slow-moving convoy ahead of him disappeared into swirling snow, it meant he was that much closer to the next phase of his life.

"Try not to be nervous when you're talking to them. Just smile and answer the questions," said his wife, Gurjeet.

Bobby glanced at Gurjeet, sitting in the passenger seat next to him. Her dark hair ran down almost to her shoulders, and she had a small duffel bag on her lap. An even larger duffel was beneath her legs, pushing her knees almost to the dashboard. It hadn't been easy to fit their entire lives into a hatchback. Amazing, how much junk they had accumulated in five years of marriage, considering they were still in their late twenties, and he was employed in one of the worst-paying professions on Earth. Actually, employed was probably too strong a word. As a matriculating postdoc, he was basically still a student. Instead of having a boss, he was technically going to work for a professor. Although this prof was so smart, it kind of scared the daylights out of him. As if packing up your life and moving to a new country wasn't terrifying enough.

"Don't be nervous? Why are you telling me not to be nervous? Why would I be nervous? We've made the border crossing a dozen times before."

Together, and separately, they had traveled from Toronto to Boston a half dozen times, beginning with Bobby's first interview at the Church Lab more than six months earlier. Then they'd made trips to find an apartment, deal with their work visas, and meet the other postdocs with whom he'd be working. He'd spent almost as much time talking to Customs agents as he had to Harvard geneticists.

"I know," Gurjeet said, trying to calm him with a smile. "I'm just saying, we both understand how it is when you get nervous — and there's no reason to be nervous."

"Of course there's no reason to be nervous."

"That's all I'm saying."

There were only two cars ahead of them now, and he could clearly make out the uniformed Customs agents standing outside the booths on either side of the highway. As the car in front reached the booths, one of the agents tapped on the driver's window, then leaned in for a brief conversation.

"Okay, but now I am getting nervous."

Then they were both laughing, because it really felt absurd. Crossing from Canada to the United States shouldn't have been a big deal. But to Bobby, Canada, especially Toronto, where he had lived his entire life, had a small-town feel. Even though Boston was technically around the same size, it felt foreign and big. Bobby's parents had left India with very little, emigrating to the UK and then finally landing in Canada. They had driven Bobby and his brother and sister to excel in school and had worked hard to assimilate into Canadian culture. Toronto was multicultural, and Bobby had never really felt like an outsider; but in America, he was going to be a foreigner. Fortunately, one visit to any lab at any university in the country, and it was easy to see that many postdocs faced the same situation.

Finally, the car ahead of them reached the border gate. After a brief talk with the Customs agent, it was waved through, and it was Bobby's turn.

He pulled the Honda to a stop in front of the gate. As the agent approached, he forced the friendliest smile he could muster, and quickly rolled down his window.

"Good morning, sir," he said, holding out his and his wife's passports for the tall agent to take.

The agent wore badges on his chest, dark sunglasses, and a gun strapped to his waist. His wide, muscular shoulders stretched the material of his uniform.

As he glanced through the first passport, a bored expression on his face, he asked, "Where are you headed?"

He couldn't have sounded less interested, and he didn't even look up from their documents, but Bobby kept the stupid grin on his face.

"Boston, sir."

The agent nodded and flicked through pages of the second passport.

"And what are you planning to do in Boston?"

"I'm going to be implanting genetically altered DNA from naked mole rats into laboratory mice, to try to reverse the aging process."

The agent stopped flicking, paused, and looked up from the passports. Bobby felt his wife jamming her fingers into his leg, and his smile started to waver. The agent stared at Bobby, glanced from Bobby to Gurjeet, then took a slight step back from the car door.

"Sir, would you please step out of the car."

Twenty minutes later, Bobby found himself seated next to his wife on a plastic-covered two-seater couch in a starkly appointed holding cell at the Canadian/U.S. border, desperately trying to explain basic genetic engineering to a room full of Customs agents. He didn't need to look at his wife's face to know that it was going to be a long drive to Boston if he ever did manage to convince the agents they weren't on their way to America to unleash some sort of genetically mutated mice on the innocent population. He had the bad habit of telling the truth — the whole truth — when he got nervous. He was a true laboratory scientist and had never been good at reading social situations, or knowing when it would be better to keep his mouth shut.

The more the Customs agents stared blankly at him, the more Bobby tried to fill the silence with explanations of the genetic anomalies that might allow naked mole rats to live thirty-year life spans, rather than the two years that regular mice lived. He just couldn't stop, even though, as he spoke, he could tell that he was just digging himself a deeper and deeper hole. They had asked him a question, so he was going to give them an answer. That's just how he was wired.

The episode reminded Bobby of the first interview he'd had with George Church, which had eventually led to him and Gurjeet packing up their lives to resettle in Boston, God and U.S. Customs willing.

Bobby's road to the Church Lab had actually been fairly smooth, up until that first interview. His parents' efforts had paid off, and Bobby had excelled in school, especially the sciences. He'd finished high school a year early, entering the University of Waterloo as one of the top engineering candidates. Originally, inspired by the movie *Back to the Future,* and hoping to one day be an inventor like Doc Brown, he'd seen himself as an engineer. Conducting pure research, with no practical aim, didn't intrigue him as much as the idea of creating something that could be used by real people.

During college, he'd married Gurjeet, whom he had met through an online Indian dating website. An emergency room nurse from a religious Sikh background, she offered a perfect counterpoint to his scientific ideologies. Even as he shifted from engineering into the biological sciences, realizing that the revolution going on in genetics had turned biology into something more akin to engineering, but with much higher stakes, she could temper his urges to push beyond reasonable boundaries.

Bobby had become fascinated by the idea of reversing the aging process. Aging had a genetic component — the way human cells declined and deteriorated with age had to do with the information coded in those twisting strips of genetic material. His personal research into the subject was what first led him to the naked mole rat.

"It's this ugly little creature," he said to

the Customs agents, getting more excited the more he talked, waving his hands in front of him. "Utterly hairless. Has no sensitivity to pain. It's completely blind. It lives its whole life underground."

Gurjeet was poking him again, trying to get him to edit himself, but he ignored her.

"But see, it doesn't get cancer. And it lives for thirty years. Regular mice live two years, three tops. But the naked mole rat is special. We don't know why. That's what we are trying to figure out."

It was the naked mole rat that had led Bobby to George Church. Nearing the end of his Ph.D. work in Canada, Bobby had begun looking for postdoc jobs that had something to do with research into reversing aging. He'd done a computer search of online job openings involving aging, by simply entering the keywords "naked mole rat." The Church Lab had immediately come up. Church and some of his postdocs had been studying the naked mole rat, to try to figure out how to transfer its amazing longevity to human cells.

When Bobby had shown the research, and the possible job opening, to Gurjeet, she had initially wondered if Church and his team were crossing an ethical line. Trying to treat aging — was this playing God? Bobby had argued back. After all, Gurjeet worked in a hospital where she changed people's fates every day. Aging, to Bobby, was a disease, and maybe Church and the naked mole rat might one day provide a cure. Bobby had instantly applied to Church's lab, and shortly afterward had been granted an interview.

At the moment, half off the couch as he went deeper into the physiology of underground rodents, Bobby could feel the sweat running down his back, just as it had when he'd first met George Church in his office on the second floor of the New Research Building. To Bobby, Church had been intimidating — tall, brilliant, not wasting any time on small talk or social niceties — and also prone to silences, like the Customs agents. The quieter Church got, the more nervous Bobby became, and he'd started to fill that silence any way he could. The night before, in order to prepare for the interview, Bobby had re-read the book Church had written about genetics, and without thinking he started to critique it — to tell Church everything he thought Church had gotten wrong. Specifically, how Church had written about DNA being the original storage medium in living cells, when Bobby believed it was most likely RNA.

When Bobby had finally finished speaking, Church had looked at him much the same way the Customs agents were now looking at him. Bobby had been about to stammer out an apology when a secretary signaled that it was time for them to head to where Bobby was supposed to give a talk about his own Ph.D. work to the lab's other postdocs.

Church walked him down the long hallway leading to the conference room in total silence. Bobby had felt as if he was on his way to the electric chair. And when he'd arrived at the conference room, things only got worse. He found himself facing a room filled with more than eighty scientists — pre- and postdocs ranging in age from their early twenties to their late sixties. He had expected to be speaking to a group of five, maybe ten people; he had interviewed at other labs over the years, and usually these things involved no more than a handful of students asking easy questions.

"My interview is open to the whole department?" he'd asked Church.

"Just the lab. Whenever you're ready."

And then Church had left him alone at the front of the room. Bobby had made it through the talk, sweating through a dozen of the most difficult questions he'd been asked since his Ph.D. thesis defense. When Church came back to gather him to return to the lab to finish his interview, Bobby tried to get a read on how he had done.

But Church had only glanced at him, then mumbled under his breath: "RNA as the original storage medium. Hmm."

That night, when Bobby had returned to Toronto, his wife had agreed — he had blown any chance of working in the Church Lab. When the email had come a week later telling him he'd gotten the position, he had been too shocked to celebrate. Then he and Gurjeet started to plan for a new life.

"And of course you're welcome to visit the lab," Bobby heard himself saying, as the Customs agents looked at him as if he was insane. "Or actually I've got my thesis work right in the trunk of the car. Just give me a moment and I'll go get it."

Their life wasn't going to be lavish, that was for sure. Postdocs were barely paid living wages, and the apartment he'd found near Fenway Park might have reminded his parents of where they had grown up in India. But he'd be working with George Church, as well as some of the most brilliant young scientists in the world. Already, he'd made an effort to meet all the postdocs he could, and to find out the various projects they were working on. Luhan Yang had stood out in particular. Her thesis work had completely blown him away, and the two of them had even hit it off. They were in the early stages of planning a project they might work on together, involving isolating stem cells from mouse ovaries and getting them to turn into mature eggs. Eventually, they hoped to do the same thing with mouse sperm; one day, it might be possible to make sex itself redundant, because it could all be done in a dish. Bobby hadn't run that one by Gurjeet yet, because he assumed she'd have some concerns. But it might have fascinating implications for the future of IVF, in vitro fertilization.

To Bobby's surprise, as he was still talking about mole rats, one of the Customs agents appeared with his Ph.D. thesis from the trunk of his car. As the other agents leafed through it, Bobby could see that they were finally beginning to believe him — he wasn't some sort of crazy genetic terrorist, he was a Harvard postdoc. Any minute now, they were going to let him go.

He couldn't wait to get back on the road. The more he thought about the Church Lab, the more excited he became. There was no end to the science he was going to be able to do there. In fact, Luhan had recently contacted him about another project Church himself had put her on, which she believed would interest Bobby. He didn't yet know the details, but he was sure that, if Church and Luhan were involved, it was going to be special.

Glancing from the agents, who were still leafing through his Ph.D. thesis, to his wife, who was ready to stab him to death with her fingernails, he guessed that, whatever Church and Luhan were up to, it probably wasn't something he should try to explain in the holding cell of a Canada/U.S. border crossing.

Chapter Seventeen

Early 2013

77 AVENUE LOUIS PASTEUR, HARVARD MEDICAL SCHOOL.

At the Elements Café, on the first floor, a circular table for four sat up against a vast picture window overlooking a highly trafficked crosswalk.

Justin Quinn, twenty-eight years old, animal enthusiast, former car salesman, community college graduate, and self-described molecular ninja, was used to feeling out of place, but even for him, this was surreal. He was at *Harvard.* In a posh little cafeteria in a medical school building dedicated to futuristic sciences, he was surrounded by people who'd mostly gone to expensive prep schools, achieved perfect SAT scores, attended top-rated colleges that Quinn could never have gotten into, let alone have afforded. Quinn had grown up dirt-poor in Amesbury, Massachusetts, raised by his mother and stepfather after his biological father walked out. He had scraped his way through higher education, taking ten years to get a college degree by way of three different universities. Yet somehow, he had found himself here, sipping black coffee from a cup with the Harvard seal inked around its base, heat seeping through the boldly scripted letters, *Veritas,* stinging the inside of his palm.

Quinn squirmed against the hard plastic chair tucked beneath the table, which was overflowing with genetic data printouts, Woolly Mammoth photos, and at least three laptops and four iPads, none of which was his. Still, he was determined to look confident, to appear, somehow, as if he *belonged.*

At the very least, he had dressed the part. He was wearing a baby blue T-shirt emblazoned with a cartoon picture of a Woolly Mammoth. The baseball hat pulled over his eyes depicted a Woolly Mammoth skeleton above the rim. And around his neck was a necklace made of real Woolly Mammoth ivory, bought from a website he'd found while scouring eBay, unsuccessfully, for a slightly used Mammoth tusk that wouldn't cost more than his childhood home.

No one could say he wasn't sartorially prepared for the first official meeting of the

Woolly Mammoth Revivalists. He felt justified in making the extra effort, considering he had the honor of being the last Revivalist to join the team.

On paper and in person, the other three members of the team sitting across from him were impressive, the brightest bunch of young people Quinn had ever met. Luhan Yang, their leader, was going to be tough to impress. And she had been looking him over with a mixture of curiosity and thinly veiled suspicion. Each swipe of her dark eyes sent chills up his spine.

Bobby Dhadwar, next to her, moving his fingers across one of the iPads, had a master's in engineering and a Ph.D. in biology, and was a little less intimidating than Luhan — he smiled a lot and his glasses were covered in fingerprints, which gave him a professorial aura well beyond his years. Quinn had heard him talking to his colleagues about how he and his wife were in the process of trying to start a family, which made him seem like a human being, at least, which was more than Quinn could say about some of the robotic-looking people he'd seen walking the halls of the New Research Building.

Margo Monroe, Luhan's other addition to the team, had a Ph.D. in biomedical engineering from BU. Smart and pleasant, she was wearing earrings shaped like elephants, which was a big plus.

Looking at the three of them, Quinn felt the need to prove his worth, to show he was ambitious, that he brought something to the table. Setting his coffee down next to one of the laptops and attacking the keyboard, he continued the conversation that Luhan had begun.

"The way I see it, we can break this down into what we're going to need, what we already have, and what we've got to do. And the first thing we're going to need are elephant cells to work with. Although the samples we will need will be small individually, over time we'll need a fair amount. Obviously we can take great care to harvest tissue without causing injury or pain, but we will need living cells. Which means contacting zoos, research centers, conservation groups."

It felt good to be talking, because at least for the moment, that meant the other team members were listening; and besides, Quinn truly enjoyed making plans. Probably because so much of early life had simply *happened* to him. As a child of blue-collar poverty, he hadn't had a lot of options. He'd developed a love for animals while volunteering with his mother at a nearby wildlife rehabilitation center, which had led to an interest in biology and endangered species, but the local public schools had offered only the most basic courses in biology: cut up a frog, look at some pond scum under a microscope.

After high school, the best option his family could afford seemed to be St. Anselm, a parochial college in Manchester, New Hampshire, run by Benedictine monks. But they'd barely made it through half the orientation presentation on the very first day of freshman year when he realized the place wasn't for him: no driving, no talking to girls after 8:00 p.m., no leaving campus. He was gone before the monks in charge started passing out name tags.

Which meant two years of community college paid for by scholarships, then a transfer to University of New Hampshire, until the money ran out. Then a few years selling cars to make enough to finish college. By the time he'd graduated with a degree in biochemistry and a minor in genetics, his mother had gotten sick — cancer — leaving him devastated and broke.

He was probably at the lowest point in his life when he met George Church, mostly by accident. Needing money, and wanting to do something in genetics and biology, he'd found a job at a start-up in Cambridge called Warp Drive. Warp Drive, one of the many companies that Church had co-founded, was focused on applying synthetic biology to organisms mined from nature for medicinal uses. The company scoured isolated places like the Amazon and the island of Rapa Nui (Easter Island), then used genetic engineering to turn natural bacteria and plants into medical treatments and cures. Quinn had found a place in the genetic engineering department, learning the ins and outs of molecular modification as he went.

So even though he didn't have the schooling of the other Revivalists at the table, he was far from a novice at genetic engineering.

"And then we need to figure out what genetic traits we're going to go after," he continued, typing into the laptop. "What makes a Mammoth a Mammoth."

The other Revivalists took over the conversation there, Luhan leading the spirited discussion while Bobby and Margo used the other laptops and iPads to conduct research on the fly. It wasn't easy, trying to pin down the most characteristic traits of the iconic, prehistoric creature. The two other laptops were opened to encyclopedia-like articles on Mammoths, while the iPads cycled through scientific journals for deeper analysis of the various possibilities.

Eventually, the team settled on thirteen traits of the Mammoth, four of which they considered most significant and would become their primary targets. The first, most visible characteristic was the thick hair for which they were named, and which protected their skin from the intense cold.

The second trait was the thick layer of subcutaneous fat that provided insulation against the severe cold of the Mammoth's habitat, and also gave them substance when they hibernated, which modern elephants didn't do.

The third characteristic was the Mammoth's small, rounded ears — very different in appearance from the large, flapping appendages of elephants.

And last, and perhaps the most difficult of the traits to isolate genetically, was the Mammoth's hemoglobin, which, unlike that of elephants and most mammals that evolved to live in temperate, or non–ice age, climates, functioned in cells at nearly freezing temperatures. For human beings — and modern elephants — in subzero conditions, the oxygen within hemoglobin binds too tightly, and the peripheral tissues freeze and die. A Mammoth's blood was able to flow and release oxygen no matter how cold it was.

The four main traits decided, Luhan shifted to the second phase of the project:

"Then comes the hard part. We need to find those traits within the Woolly Mammoth's genetic sequence. Figure out which genes code for those traits."

"Which means we need a properly sequenced genome of a Mammoth," Bobby chimed in.

"What about the Penn State project?" Margo asked. Penn State had conducted the original Mammoth sequencing that had led Nicholas Wade from the *New York Times* to call Church — the reason they were all gathered together in the first place.

"Funny thing about that," Bobby said. "They actually had a partial sequence up online for a while. But they took it down. I tried contacting them, but they haven't gotten back to me."

Most likely, Quinn guessed, there were problems with their sequence. It was a pretty incredible task — sequencing a ten-thousand-year-old creature from a small sample. Not to mention the fact that the Mammoth had four billion base pairs in its

DNA, a full billion more than the DNA of a human being.

"There are other groups with Mammoth material," Luhan said. "There's a team in Chicago working on it, and also the Reich lab here at Harvard. The Reich lab's sequence should be of sufficient quality, and Dr. Church mentioned to me in our last meeting that they'll be open to letting us use it for our Revival."

There were certainly benefits to being at Harvard, and in Church's lab. In fact, Church had recently gotten them their own sample of frozen Mammoth from a Canadian collector, a little piece of Woolly intestine that was now sitting in a freezer on the second floor of the New Research Building. If they'd had the time, the money, and the inclination, they could have sequenced it themselves; but fortunately, they wouldn't need to. Because Church was one of the biggest proponents of open-source science, he had fostered a wonderful sense of camaraderie among other genetic scientists.

It was Church's openness that had led to Quinn's having the opportunity to be at that table in the first place. Once Church, Brand, and Phelan had officially decided to launch the Revival project, they had arranged a TEDx Talk on de-extinction, in conjunction with *National Geographic.* Quinn had listened to it multiple times, and had immediately wanted to get involved.

As a cofounder of Warp Drive, Church visited the start-up about once a month; it was during one of those visits that Quinn had jumped all over him.

"Dr. Church, I need to do this. I've dreamed of doing this. It will validate my existence."

Maybe it had been over-the-top, but Church had been impressed enough to put him in touch with Brand and Phelan, and their phone discussion blossomed into an invitation to join the Revivalists.

"Once we've got the genome of a mammoth," Luhan continued, "we need to find the genes that code for the traits we've chosen. Hair, ears, subcutaneous fat, and hemoglobin."

As complicated as that sounded, it was one of their simpler tasks. The key to identifying the four traits actually lay in the Human Genome Project. Since the HGP had been completed back in 2003, scientists had moved on to identifying the roughly twenty thousand specific sequences of genes amid the over three billion base pairs of DNA that coded for traits in humans, from eye color to various inherited diseases and conditions.

So, if the team wanted to figure out a specific gene in the Woolly Mammoth sequence — say, red hair — they would choose a similar trait in the human genome that was already sequenced, and then use a computer matching program to isolate the similar sequence within the Woolly Mammoth genome. It was as easy as plugging parameters into a search engine: Google DNA.

Hair, ears, subcutaneous fat, hemoglobin — they were all searchable genes, as soon as they had their Woolly Mammoth genome. Tusks, among the other traits they hoped to work on later in the project, would be more difficult. Humans don't have tusks, so locating DNA coding for tusks in a Mammoth would take guesswork and trial and error.

"Once we've figured out the genes we need," Luhan continued, "we begin the process of synthesizing them."

They all understood that they weren't going to *clone* a Woolly Mammoth. They were going to *make* one. They weren't going to transfer genetic material from a frozen carcass, they were going to create the material in a dish and implant it within a living elephant cell.

It was synthetic biology, which Quinn did

every day at Warp Drive. Unfortunately, that real-world experience hadn't been enough to get him into graduate school; without the right college pedigree and the right recommendations, he had been rejected wherever he'd applied. The lab did have non-Ph.D.s, but Quinn wasn't a Harvard employee, so, technically, he wasn't allowed to work in the Church Lab, and he had to be brought in for the Revivalist meetings as a guest of one of the postdocs. But his hands-on experience made him an important, capable part of the team.

"And then we place the genes into the elephant cell," Quinn said, typing the fourth and final stage of their work into the laptop.

That was the magic act, the transformation in genetics that George Church was leading — *no longer reading, but writing DNA.*

Quinn could feel Luhan looking him over, and he knew what she was thinking. This fourth stage of their project would be cutting-edge science, with the emphasis on *cutting.* Because the process by which they would place those genes into the elephant's living cell represented one of the most important developments in modern science.

CRISPR — "clustered regularly interspaced short palindromic repeats" — was a scientific breakthrough that revolutionized genetic biology. A new method of editing large numbers of genes simultaneously, it had jumped onto the scene just six months earlier, through a pair of papers published in *Science* magazine, one of which had come from George Church and the Church Lab itself (including Luhan), plus three other papers in the same month. Multiple labs claimed credit for having invented CRISPR.

Luhan, Bobby, and Margo had come of age as scientists using CRISPR, and they understood it as well as anyone. Quinn worked with CRISPR every day, and in fact often volunteered to speak at local high schools, because he knew, as complex as the science behind CRISPR might be, it was such an important tool that everyone should have at least a basic understanding of how it worked.

Over the past few months, he had tried to come up with simple ways to explain what was essentially now the primary tool in genetic engineering.

With high school students, he started with the basics. Every cell in every living creature contained a copy of the creature's genome — the double strands of DNA, made up of billions of base pairs of molecules, that coded for every trait or characteristic that made the creature what it was. Those double strands, known as the double helix, the discovery made by James Watson, Francis Crick, and Rosalind Franklin in the fifties, were connected in pairs of chemical bases: the chemical adenine paired with thymine (A-T) and the chemical cytosine paired with guanine (C-G).

Simple bacteria had developed a natural system to protect their own genomes — their double strands of DNA — from attacks by viruses. To defend themselves, they employed a protein called Cas 9 that could act like a molecular pair of scissors. When a virus attacked, spreading its genetic material into the bacteria, the Cas 9 protein, guided by a string of messenger RNA — a single strand of nucleotides that carries information from DNA and uses that information to assemble proteins — that would match up with the invading DNA, would slice the virus's genetic material at precise points in its connected pairs. The RNA essentially became a targeting system, allowing the Cas 9 to slice and destroy invading viruses.

Six months ago, a few teams of scientists across the globe — including Jennifer Doudna, Martin Jinek, Feng Zhang, Le Cong, and Prashant Mali, as well as Luhan in Church's lab — realized that the Cas 9 protein could also be used to cut strands of DNA from other organisms besides viruses. They could use synthetically created guide RNA to aim Cas 9 at any chosen sequence of genes, slice them out, and then replace them with a different sequence, created in a lab. In his paper in *Science,* Church had demonstrated for the first time that the CRISPR technique could be successfully used in human cells.

In short, placing Woolly Mammoth genes into elephant cells was as simple as designing a guiding segment of RNA to match the base pairs at the ends of the sequence that coded for whatever trait you were hoping to replace — and the Cas 9 would do the rest. The RNA would line up the synthetic gene with the similar elephant gene, slice the DNA with the Cas 9 molecular scissors, and the genome itself would reattach the new, Woolly Mammoth gene in its place, through a natural healing process.

The elephant cell would no longer contain an elephant gene. Instead, it would contain a synthetic Woolly Mammoth gene, coding for the desired Mammoth trait.

"And when that succeeds — when we've proven our ability to insert a Woolly Mammoth gene into an elephant cell — we shift our work to the stem cells."

Stem cells are the undifferentiated cells within every living thing that can give rise to the diversity of other cells that make up the living whole. Most cells are specialized and don't change that specialty — the cells making up an ear or a hair remain that of an ear or hair. But stem cells can differentiate to become anything: ears, hair, heart, lungs, hemoglobin, tusks. A genetically changed stem cell could give rise to all the different traits of a Woolly Mammoth. And those traits would then be inheritable and inherited. The new creature wouldn't just look like a Woolly Mammoth and behave like one. It would *be* a Woolly Mammoth.

At that point in the meeting, Luhan took total control, splitting up the team in order to take on the various tasks. Bobby and Margo would contact the zoos in search of elephant samples. Luhan would do her best to obtain a complete Woolly Mammoth genome, and then go through the genetic sequence searching for the specific traits they wanted to replicate. And Quinn would begin to design and prepare the CRISPR components they would need to implant the synthesized Woolly Mammoth mutations into the elephant genome.

By the time they were ready to leave the

café, their excitement was as keen as a pair of molecular scissors.

"Just remember," Bobby said, smiling, as he gathered up the laptops, "it's only science fiction until we remove the fiction. Then it becomes real."

Chapter Eighteen

Early 2013

Two days later, Bobby wasn't smiling after cold-calling zoos and conservation parks, trying to get access to elephant tissue. He held a desktop phone against his ear and listened to the braying of the zoo manager on the other end of the line. Bobby was hoping to get a word in before the manager worked himself past all reasoning, but the guy wasn't even pausing to breathe, his words were just running into one another like a train jumped free from its tracks.

"No," he said, finally breaking into the frenzied zoo manager's monologue. "We aren't going to clone your elephants. We just need a little sample for our experiments . . .

"No, we aren't going to kill any elephants. Why would we want to kill elephants . . . ?

"Yes, I have seen that movie. Yes, the dinosaurs were very frightening . . .

"Yes, I've seen that movie, too. No, we

aren't trying to make scary mutant animals . . ."

He considered hanging up, but he'd already introduced himself as a member of the Church Lab and didn't want to appear rude. He felt a certain level of responsibility — especially since he was sitting in Church's office, in Church's chair.

It hadn't been his choice to make the calls from there; he'd have been perfectly happy using his cell phone in the café. But Luhan had convinced him he'd sound more official if there wasn't a din of background chatter behind him. And besides, Church wouldn't be returning to the office anytime soon.

At the moment, Church was off putting out a huge public relations fire, tangentially related to the Woolly Mammoth Revival project and his work on de-extinction. As Bobby understood it, the problem arose from an interview Church had given to the German magazine *Der Spiegel* back in January. The interviewer had taken note of a passage in Church's most recent book, *Regenesis: How Synthetic Biology Will Reinvent Nature and Ourselves,* which had laid out the idea of de-extinction and the science that would allow them to revive the Woolly Mammoth, as well as other extinct species. One of those species Church had mused about in the book was the Neanderthal. After explaining the steps of genetic engineering necessary to revive the Woolly Mammoth, the passage read:

> The same technique would work for the Neanderthal, except that you'd start with a stem cell genome from a human adult and gradually reverse engineer it into the Neanderthal genome or a reasonably close equivalent. . . . If society becomes comfortable with cloning and sees value in true human diversity, then the whole Neanderthal creature itself could be cloned by a surrogate mother chimp — or by an extremely adventurous female human.

After the interview was published, it was distorted, in Church's view, by the *MIT Technology Review* to make it seem as if Church had already moved well beyond theory in his work — and the *Review* essentially placed a "Want Ad" for women who would carry to term a Neanderthal baby. Church was at a Martin Luther King's Day barbecue when his cell phone rang. He picked up to hear the dean of Harvard Medical School on the other end.

"Did you put out a call asking for women to volunteer to carry a Neanderthal baby?"

the dean asked.

The situation would have been humorous, if it hadn't caused such an uproar. Church had been forced to contact numerous newspapers and magazines, including the *Boston Herald,* to explain that he was not, in fact, in the process of bringing back the Neanderthal species. Even more bizarre, the Church Lab had received hundreds of letters from women volunteering to carry the first Neanderthal baby, proving two things. There were plenty of "extremely adventurous female humans" out there in the Greater Boston area. And genetics was a powerful tool, but also an ethical minefield. It was incredibly easy for a scientist to be misunderstood when speaking to the general public. No matter how well-meaning a scientist is, there were always going to be people imagining the worst.

Bobby was learning that lesson again, firsthand.

Every call had been the same. At first, the zoo managers were respectful, impressed by the Harvard name. But as soon as Bobby mentioned genetic engineering, their attitudes changed to suspicion, fear, even anger. Everyone he talked to jumped to the conclusion that he was trying either to clone their elephants or turn them into some sort of genetic monster. No matter what Bobby told them, they kept turning him down.

"We just need a little sample," he said, trying again. "The elephant won't even notice." Bobby may have been minimizing the process a bit; the elephant would certainly notice when they harvested a small amount of tissue, but it wouldn't cause the animal any pain, or have any lasting effect.

The zookeeper did not want to listen. Bobby sighed, thanked him for his time, and hung up. He crossed out another name on the list. He was getting dangerously near the bottom of the page and considered giving up; perhaps they could find some frozen elephant material at another lab, or at some culture bank at a university. But Luhan would say that it was better to have a live, fresh cell. To someone as determined as Luhan, it wouldn't matter how many people had said no. It only mattered that one said yes.

Bobby shifted to the next entry on the list, a private zoo in a suburban area about forty miles west of Boston. According to Bobby's notes, they had a four-year-old pair of healthy young African elephants.

He didn't expect much as he dialed the number, but when a kindly-sounding older gentleman answered the phone, he put on his most official-sounding voice and started into his appeal, honed by forty hours of straight rejection. He explained the Woolly Mammoth Revival project, then asked for the tiniest of cell samples.

To his surprise, he didn't get an immediate no. Instead, the man told him he was welcome to visit the zoo and take a look at the animals for himself. Then the man added: "Considering what you're trying to do, I don't think you've really thought this through. But I'm happy to help if I can."

Bobby was too excited to try to figure out what the man's cryptic warning meant.

Step one of the Woolly Mammoth Revival, completed: They'd found their elephant cells.

Two days later, Bobby joined Luhan, Quinn, and Margo at their regular window table at Elements Café, and the four of them gathered around one of the laptop computers. The screen was open to a video shot from Bobby's cell phone, taken not fourteen hours earlier. His phone had been held close enough to the chain-link fence surrounding the elephant habitat at the private zoo that most of the enclosure was visible. A grassy pasture led up to a kidney-shaped, man-made lake, which was surrounded by a few bushes and the kind of low trees found on a savannah. For a few seconds, they saw nothing else, and then, down the center of the screen, a middle-aged man came running as fast as he could toward the fence. One hand held on to his hat, the other clutched a large hypodermic needle. His face was contorted in fear. Dirt kicked up from the heels of his boots as he plunged down the center of the pasture, heading for a pair of barricades on a steel secondary enclosure that ran just inside the chain-link fence.

He'd almost made it to the barricades when the elephant pursuing him came into view. Onscreen, it looked twice as large as Bobby had remembered, and it was moving fast, almost as fast as the zoo manager. Its huge, floppy ears were pinned back against the sides of its head, its tusks were raised, and its trunk was curled up into the air, a jagged semaphore. The look in its eyes was pure anger, and the entire zoo — and the camera — seemed to shake as it charged forward.

A few seconds later, the zoo manager made it through the barricade and the chain-link fence, gasping as he locked everything shut behind him. The elephant skidded to a stop a few feet from the barricade, then pawed at the ground with an enormous padded foot. Finally, it gave a shake of its massive head, turned, and started back in the direction it had come.

As the computer screen went black, Luhan was the first to speak.

"African elephants seem to be a bit aggressive."

Quinn coughed. "You think? He didn't even come close to getting a sample in that needle."

"We really don't want to turn one of them into a Woolly Mammoth," Bobby said, his voice dejected.

None of them needed to complete his thought. They had all seen *Jurassic Park*. Rampaging Woolly Mammoths tearing up the Siberian countryside were not going to win any of them a Nobel Prize.

"So what do we do?" Margo asked.

There was a brief pause, and then Luhan shrugged.

"We pivot."

From a genetic standpoint, African elephants had never been their first choice. As a species, Asian elephants were much less aggressive and temperamental, and so were also a much better fit for their project. Genetically, they had more in common with the Woolly Mammoth than the African species and thus were a closer relation to the

Mammoth, which would make the actual sequence engineering a little easier. This made sense, since the Asian elephant came from the same continent where the Woolly Mammoth had lived in its largest populations.

Luhan had made progress obtaining the sequencing. The Reich lab at Harvard had kept its word, and the Revivalists had gained access to the beginnings of an okay sequence of the Mammoth genome. Luhan had already begun running the necessary search programs to find matches for the traits they were looking for: hair, ears, subcutaneous fat, and hemoglobin.

Soon, they were going to need their elephant samples, and Asian elephants were more difficult to find. An endangered species, Asian elephants' wild population was fewer than fifty thousand animals. Their population had declined by at least 50 percent over the past three generations, approximately sixty to seventy-five years.

"They're not in many zoos," Quinn said, picking up on the rest of the team's thoughts. "If zookeepers gave Bobby a hard time about giving us access to African elephants, they're going to be even tougher letting us get close to an endangered species."

They were all quiet for a minute as they thought, the ambient sounds of the café behind them reverberating. Finally, Luhan broke the silence.

"Well, we don't need the *whole* elephant," she said.

"What do you mean?" Margo asked.

"Our eventual goal is stem cells, right?"

In order to implant a genetic change — Mammoth DNA — into the elephant spe-cies DNA, they needed a stem cell. Origi-nally, that was going to be something they would seek after they'd experimented on living elephant cells.

"Sure," Bobby said. "But it's hard enough getting live elephant samples. Where do we find elephant stem cells?"

"Elephant placenta," Luhan responded.

Mammalian placenta is rich in stem cells, and placental cells were as young as fetal cells, thus easier to reprogram than adult somatic cells (any biological cell that forms an animal's body). It was the reason some human couples banked placental tissue after their child's birth — to freeze stem cells for use in later medical procedures, such as treating leukemia, various cancers such as Hodgkin's lymphoma, and nearly eighty other childhood and early adult diseases.

"You've got an elephant placenta lying

around?" Quinn asked.

Luhan smiled, briefly.

"The next best thing."

She turned the laptop toward her, typed in the keywords for another website, and brought up a new video, a live feed of another zoo enclosure. Near the back of the screen, they could just make out what appeared to be a female Asian elephant, her swollen abdomen almost reaching the ground. She was so heavy that it looked as if her knees were about to buckle.

"She's pregnant," he said, quietly.

"It's a live cam," Luhan explained. "Out of Chicago. It's on twenty-four hours a day. I've been watching it on and off for most of the week."

"Christ," Quinn said. "How long has this been running?"

"I don't know. An elephant gestates for twenty-two months."

Elephants have one of the longest gestation periods of all animals. Eventually, this would become a problem for the Woolly Mammoth Revival — because no matter how quickly they could sequence, synthesize, and implant a gene into a fertilized egg, it would take almost two years for an elephant to give birth.

"So there are people who actually sit

around watching a pregnant elephant online?" Quinn asked.

"There's a television channel in the Philippines that just shows an aquarium all day and night," Bobby said. "People are strange."

"What this means," Luhan interrupted, "is the minute this elephant gives birth, we'll know. If we can get the zoo to agree, we can get someone to the zoo to get a piece of the placenta. Fresh and ready for use."

Bobby thought for a moment.

"It will only be fresh if we act quickly," he said. "Which means we'll have to know right when she goes into labor. We'll have to be watching this all the time."

"I think I might know someone in Chicago who can help us out," Quinn said. "Won't cost us much more than the cheapest ticket we can find from O'Hare to Logan."

Luhan nodded. Quinn was growing on her. She glanced around the table, at each member of the team, then pointed back at the screen.

"So who wants to take the first shift?"

Chapter Nineteen

April 2013

Jack Walton had done some pretty strange things to make money in his twenty-eight years of life, and he had put up with a lot of crap to earn a few extra dollars along the way. But as a higher-education lifer — a grad student in the ninth year of what felt like a never-ending Ph.D. — he couldn't be particularly choosy. He'd worked the door at a strip club in Elgin, Illinois, during his third year in the biology department at U of Chicago. He had locked himself in one of the bathroom stalls while the girls were changing shifts to catch a few extra minutes of work on a paper on unique protein inhibitors found in E. coli bacteria. He'd driven an ice cream truck over two summers, stacking issues of *Science* and *Nature* next to the freezer filled with rainbow pops and SpongeBob SquarePants ice cream sandwiches. Just last winter, his fingers chilled to the bone by thirty-mile-per-hour winds and snow that fell in clumps the size of dying angels, he'd sharpened skates and helped tourists lace them up before they tried their luck on a frozen chunk of Lake Michigan.

But none of the jobs he'd had before had been as absurd as watching an elephant give birth. He'd never done it before — and he damn well never wanted to again.

Jack was standing near the outer edge of the thick glass enclosure that surrounded the pregnant Asian elephant's habitat, leaning against a shoulder-high railing. A large white plastic medical-grade cooler was on the ground between his boots. Next to him was one of the zoo hands, a kid barely out of his teens, with curly blond hair and a goofy smile on his face.

"Incredible, isn't it," the kid was saying. "The circle of life."

Walton didn't see anything circular going on, just a lot of blood and mucus and a poor, bulbous elephant crouching uncomfortably on her back legs as the amniotic sac began descending out of her, like a balloon escaping through a valve four sizes too small.

Walton cringed as the sac inched downward, and the elephant raised her trunk in obvious pain. A half dozen zookeepers surrounded her, coaxing her along. Hundreds more people were watching via the live-feed camera trained on the elephant from near the ceiling of the enclosure. Walton guessed that for a lot of people, like the kid standing next to him, this was a miracle in the making, but to him it was pretty hard to stomach. Then again, there was a reason he'd chosen academia rather than medicine like both of his brothers. He liked the things that were supposed to be inside to stay inside.

Then again, as the curled-up shape of the baby elephant suddenly became visible, the legs pressing out against the whitish bubble of the sac, he felt a rush of awe. The feeling was short-lived, as a second later, the baby and sac burst out of the mother elephant with the force of a Jacuzzi jet, splashing down to the ground. A stream of thick reddish liquid followed, and Walton quickly averted his eyes as the zoo kid slapped his back.

"Dude, how cool was that?"

"It was something. I assume that's the placenta?"

Walton turned back toward the enclosure and watched as the mother elephant poked the baby with her trunk and front foot.

There was a nerve-racking moment before the baby started moving — then it opened its mouth and made a sound not unlike an infant's cry.

"That's right," the kid said. "Keep watching, pretty soon he'll try to stand up. It's an amazing thing to see . . ."

"The placenta," Walton repeated. "Can we go get it now?"

The kid looked at him in disbelief. "Are you kidding?" he asked.

As Walton watched the mother hovering over the baby, he noticed that the other zookeepers were keeping their distance as they clapped their hands in joy at what they'd just seen. No matter how excited they were, none of the professionals was moving anywhere near this mother and child.

"I need it fresh. They were pretty explicit," said Walton.

When Justin Quinn had first explained the job to Walton a few days earlier, he had thought it was a joke. Retrieving an elephant placenta and carting it to Boston was bizarre enough, but the reason Quinn and his team wanted the sample was a story Walton would be retelling for years to come. Here he was, toiling away for years on E. coli, and Quinn was playing God with a freaking Woolly Mammoth.

Sure, it was probably pure fantasy. Walton had met Quinn only once, at a synthetic biology conference in a convention center outside Boston. The kid had struck him as whip smart and superambitious, but he didn't even have a Ph.D.

They'd gone out drinking after the conference at some dive bar Quinn knew about in South Boston. They'd met a couple of real winners, a blonde and a brunette who were just bored enough with the local male talent to pretend to be impressed by a pair of biologists who could recite the periodic table by heart. Hell, if Quinn had told his Woolly Mammoth story that night, they might have actually scored, rather than ending up two hours later at a diner by themselves.

Still, when Quinn had called him a few days ago with the strange job offer, Walton hadn't even considered turning him down. Two hundred bucks and a round-trip flight to Boston was nothing to sneeze at, considering he had maxed out his credit cards on last month's rent.

"Fresh or not," the zoo kid said, "we're not getting anywhere near that placenta for a while."

"Why not? I thought Asian elephants were

the friendly ones."

"Asian elephants are usually pretty docile. I'd even call them friendly. But when an Asian elephant gives birth, that's another story. They get very protective."

Walton raised his eyebrows.

"Even of their placenta?"

The kid shrugged.

"You go in there right now and mess with that elephant, I can't promise that you're going to come back out."

Walton winced. That sounded pretty definitive.

"So we wait. How long?"

The kid gestured toward the mother elephant.

"Ask her."

Christ. Walton could have chosen to work at a real job for a living. His brothers had nice houses and played golf on the weekends. But he had chosen to be a scientist.

He stood around elephant enclosures with a cooler, waiting to get his hands on some placenta.

Eight hours later, Walton found himself sitting at a bar in a domestic terminal at O'Hare, making eyes at a very pretty flight attendant at a table four feet away. He'd sent her two drinks already, the second of which she was nursing as she talked on her cell phone. She'd smiled at him when the first drink arrived, but since then had pretty much ignored his existence. He was about to write it off as another failed experiment, when she finally put down her phone, downed the rest of drink number two, and slid onto the barstool next to him.

She introduced herself, Marty something, a pretty name to go with her face. He was in the middle of telling her about himself — well, leaving out the near poverty and the fact that he still lived in a dorm room at twenty-eight — when she glanced down below his feet and saw the huge plastic cooler.

"You packed refreshments for the flight?"

"Not exactly."

"So what's inside?"

He thought about making something up. He remembered how one of his brothers used to do transplant runs during med school: He could say he was carrying a heart or a liver — she might have been impressed by that. But for once, he decided the truth was compelling enough.

"Elephant placenta," he said.

When her eyes widened, he grinned.

"Wait until I tell you what they're going to do with it."

Chapter Twenty

April 2013

400 TECHNOLOGY SQUARE, CAMBRIDGE, MASSACHUSETTS, WARP DRIVE LLC, BIOSAFETY LEVEL TWO REVERSE AIRFLOW LAB, THIRD FLOOR.

You start off in a sterile changing room. Remove your civilian clothes, fold them up, and jam them into your locker next to your coat and shoes. Then you put on the scrubs. Light blue, elastic band at the waist. Transparent plastic cover over your hair. Latex gloves over your hands, tightened at the wrist. White mask over your mouth and nose.

Five minutes to get through the first air lock. Ten minutes in the reverse airflow chamber, airborne contaminants filtered through giant vents in the ceiling and floor. Then into the second air lock.

Finally, you step into a sterile, shiny lab. Everything is white: the cement walls, the smooth, polished floor, the equipment shelves. A glass metropolis of test tubes, beakers, specimen jars, and measuring vessels lined up in rows.

You head for one of the biosafety cabinets, brand-new, lined in stainless steel, flush with one of the walls. You sit beneath the flow hood, facing a counter partially covered by a glass sash.

Lined up across the cabinet, your Petri dishes. Tissue cultures, each one containing living Asian elephant cells. The culture medium in the dishes has a pinkish hue.

The first thing you need to do is express an enzyme — Cas 9, your CRISPR construct. And then you need to feed that enzyme a length of guide RNA. The guide RNA will line the Cas 9 with a specified length of DNA in the elephant cells, like a Band-Aid placed along a skin gash. You'll place the mixture into a gel electrophoresis unit, hit a button, force a heavy voltage to run through the material to cause the cell membrane to become polarized, allow the DNA inside. Then the Cas 9 will do the rest, slicing through the base pairs at the ends like a microscopic, molecular pair of scissors and inserting the new but ancient DNA into the elephant DNA . . .

Quinn leaned back in his chair, looking down into the Petri dish, his gloved hands inches above the surface, a tiny pipette hanging from his fingers. The sweat was beading across his forehead and down the back of his neck. It wasn't just that he had sneaked into his company's headquarters after hours to work on the Revival project without authorization from his superiors (or from George Church). No matter how practiced he was, this was painstaking, difficult, anxiety-inducing work. After he inserted the CRISPR enzymes, it would be two days before he saw the results. But he was confident that at the very least, they would have proof of concept: They would have inserted a sequence of synthetic Woolly Mammoth genes into an elephant cell.

And yet Quinn also knew that all of this painstaking work was ultimately futile. Because the cells in the Petri dish in front of him, suspended in the pinkish medium, were not stem cells. Though the placental tissue that his friend from Chicago had provided them had indeed been fresh, they had been unable to generate workable stem cells from the material, no matter how much they harvested.

A quick search of the literature told them that they were not alone in this failure: It turned out that nobody had successfully generated and manipulated elephant stem cells yet. Mouse stem cells, pig stem cells, even human stem cells had been used in genetic engineering experiments, but never elephant stem cells.

Luhan had a theory about that. It was widely known that elephants did not get cancer. It was something that researchers couldn't yet explain — an animal so big, with so many cells replicating at such a rate for so many years, providing so many opportunities for misfires, mutations — elephants should, as a species, be rife with cancer. But cancer in the great gray mammals was exceedingly rare.

Luhan believed that the same mechanisms that protected the elephant from cancer might be linked to the difficulty in generating their stem cells. Whatever the reason was, it was presenting an enormous roadblock in their work. Not only did they need stem cells if they were eventually going to get the genetic traits of the Woolly Mammoth to express themselves in a growing embryo, there also were finite limits to how much they could experiment with nonstem elephant cells, whether they were skin, blood, or placenta. Living cells, no matter what animal they came from, could divide only a specific number of times before the telomeres that held them together withered and broke, and no longer provided the necessary protection to the genetic material. In human cells, it was known as the "Hayflick limit": Human fetal cells could divide and replicate only a certain number of times — between forty and sixty — before they stopped dividing and aged. Elephant cells had the same limitation.

Stem cells, on the other hand — human, elephant, and other animals — did not fall under the Hayflick limit. They could divide — and survive — indefinitely. In fact, it was one of their two main properties: They could self-renew, and were thus immortal. And they were "pluripotent" — which meant they could turn into any other kind of cell, giving rise to everything from skin, to hair, to hemoglobin.

Which meant that, without stem cells, the Woolly Mammoth Revivalists would never be able to see if their genetic implantation was actually working; their cells would die before any visible traits could be expressed.

And yet, alone in the lab in Church's start-up in Cambridge, Quinn didn't feel frustrated. Quite the opposite, he was elated. And it wasn't simply the fact that he had sneaked the work into Warp Drive — it wasn't a safety issue, just bureaucracy — and wasn't really supposed to be part of the

Church Lab at all that was sending adrenaline through his veins.

He was doing real, frontline science. In that pipette was a string of DNA that hadn't existed, alive, in thousands of years. Even though it would end up little more than a tiny shift in a single elephant cell trapped in a Petri dish, it was a momentous moment. The fact that it was, so far, a failure in the greater scheme of things meant very little. Failure was an important part of the scientific process. It was failure that spurred science toward innovation.

This was where they were: They could put their synthetic genes into elephant cells.

It was a start, but it wasn't enough.

Luhan, Bobby, and Margo were feeling the same mixture of frustration and delight, and asking themselves the same questions.

Had they hit a wall they couldn't breach?

Or was there some other way to go?

■ ■ ■ ■

PART FOUR

■ ■ ■ ■

There's a lot of faith expressed by scientists about science. It's kind of an act of faith that science is a good thing. We don't know that for sure. We may not know that millions of years from now.

— GEORGE M. CHURCH

You can't just hoard your ideas inside the ivory tower. You have to get them out into the world.

— GEORGE M. CHURCH

People think it's great to be ahead of your time, but it can actually be quite painful.

— GEORGE M. CHURCH

Chapter Twenty-One

May 2013

MUUS KHAYA, SAKHA REPUBLIC, SIBERIA, ELEVATION — PARTWAY TO A PEAK OF 9,708 FEET.

Three hundred miles south of the Arctic Circle.

Geneticist Jy Minh, senior scientist of the Sooam Biotech Research Foundation out of Seoul, South Korea, crouched low behind the row of sandbags at the edge of the natural ice cave. He pressed his hands tight over his ears, waiting for the first charge to go off. No matter how prepared he was for the moment, no matter how carefully he had supervised the placement of the tapered explosive pills, how painstakingly he'd checked and rechecked the detonation caps, he couldn't keep his mind from reviewing all the things that could go wrong. What if he'd miscalculated the strength of the chemical explosive, or the sturdiness of the cave roof? What if he'd set the charges in the wrong place, or had missed something on the ultrasound readings they'd made of the cave wall? What if, somehow, he was about to blow them all to hell?

Of course, his thoughts were absurd. The preparations for this moment had taken weeks and had involved dozens of scientists from South Korea and Russia. Minh himself had spent most of the past ten days climbing up and down the winding trail from their base camp, situated in a small clearing halfway down Muus Khaya, checking and rechecking his calculations. Today alone, he'd been in the ice cave since five in the morning, ignoring the growing numbness in his fingers and toes as he readied his team for the final moments of the excavation project.

Minh hated the cold, and he couldn't wait to get back to the relative comfort of camp. A canvas tent fortified with yak fur, as desperate as it seemed, was leagues better than a frozen hole carved into the face of one of the northernmost mountains in the world. Minh could tell, by the graying sky peeking through the heavy clouds that dominated the view from the cave's opening, that it would already be getting dark by the time he made his way down to camp.

And the temperature would drop even further, maybe ten or twenty degrees, which would slow his descent to a crawl, even with the assistance of the pair of Yakut hunters he'd added to his team just days ago.

Minh didn't relish the idea of climbing down in the dark, but the idea of spending the night in the ice cave was even more distasteful. A scientist, a geneticist by training, Minh certainly didn't believe in ghosts. But he couldn't help thinking that if apparitions did exist, this was exactly where you'd find them.

"Ten seconds," one of the demolition experts from his team called out, from behind another wall of sandbags a few feet closer to the ice wall. "Base camp has been notified, the director has given the go-ahead. Beginning countdown."

"Ten. Nine. Eight."

Minh hunched lower, his face just inches from the solid, blue-tinged floor of the cave. Despite the cold, the place was physically beautiful. Minh could still remember the moment when he'd first made the climb up the mountain and set his eyes on the natural geological formation, taking in the sheer walls and floor, the arched ceiling nearly twenty feet above. Aside from a sprouting of stalagmites near the inside corner of the cavern, and a pile of debris from a past avalanche that had to be cleared out, partially obscuring the opening of the cave from outside view, the cave had looked like this — pristine, frozen, timeless — for tens of thousands of years. It was a perfectly preserved time machine just waiting for his team to stumble through the yawn of the cave's mouth.

A cave halfway up Muus Khaya had not been Minh's or his team's first choice for their expedition. The task of moving equipment up and down the rough trail to base camp — let alone transporting whatever specimens they'd manage to recover, via all-terrain vehicles, to the airfield in Teply Klyuch, for the trip back to Seoul — made it a logistical nightmare. Their original plan had been simpler and more elegant — and it had worked. With the assistance of the Russian military, they'd taken boats down the Yana River and used enormous fire hoses to blast holes in the icy cliff faces that lined the banks. They'd made some incredible finds, gathering many wonderful specimens for the foundation's labs. Minh would have been content with what the hoses and man-made caves had provided, until he heard the stories from the handful of Yakut hunters they'd hired to help them along the

Yana: stories of a natural cave, containing specimens much better preserved than anything they'd retrieved by blasting ice with their hoses.

That first day up Muus Khaya, Minh had learned to trust the Yakuts. Minutes after entering the natural ice cave, he'd found his first specimens: two Siberian cave lion cubs, in nearly perfect condition, buried under only a few feet of clear, bluish ice. More than twelve thousand years old, and yet so untouched and undamaged that it seemed they might spring up onto their feet at any moment. Thrilled to his core, Minh had managed to put off celebrating long enough to bring in the portable ultrasound devices and begin scanning the floor and walls for potential other finds. And that's when he'd made an even bigger discovery.

"Seven."

"Six."

"Five."

Minh began to tremble, resisting the urge to raise his head above the sandbags, to check the charges one last time. Even without looking, he could picture them, positioned in a small ring affixed to a section of the back wall of the cave, right against the floor. Just enough explosives to blast away a cone-shaped section of the ice, to a depth of precisely nine feet.

"Four."

"Three."

"Two."

It wasn't ancient lion cubs that had brought Minh to the Arctic, or had sent his team up and down the Yana blasting caves into the ice. To be sure, the cubs were a fantastic find — they had already been sent back to the labs at the foundation, and the scientists had already begun harvesting the extinct animals' well-preserved cells. But Minh's superiors in Seoul had a much larger target in mind than simply attempting to revive an extinct species.

As a senior geneticist at the foundation for the past five years, Minh understood, better than most, how far his company's cloning technology had advanced since the scandal that had nearly shut down the company more than ten years ago. Sooam itself was best known for its business of cloning dogs for profit, charging clients, mostly wealthy Americans, one hundred thousand dollars a pop to bring beloved pets back to life via a cloning factory that could turn a skin sample into a viable canine embryo, but the foundation had advanced the art of cloning to include many large mammals for altruistic purposes, from cloned pigs that might one day aid in the cure of diabetes to cloned cows that could solve the world's hunger problems. And yet, no matter how far the foundation had come — leading the world in cloning science — it couldn't wash away the stain from a decade ago.

Minh could still remember the photos from the newspapers, days after the foundation's founder, Hwang Woo Suk, had been expelled from his professorship at Seoul National University, his laboratory raided by government agents, his awards and accolades rescinded. Just a few years earlier, Hwang had risen to international fame by publishing a couple of papers in *Science,* announcing that he had cloned the first human embryo, creating a stem cell line that could be used for everything from curing disease to growing organs for transplant. But shortly after the announcement, rumors began to circulate that Hwang had reached this achievement using eggs donated by his own grad students — that in fact, he had personally brought at least one student to the harvesting table — raising enormous ethical questions about the voluntary nature of those donations. Even worse, most of the data in his two articles allegedly were falsified. More than just a simple fudging of numbers, it appeared that Hwang's team hadn't successfully cloned human embryos at all.

Facing jail time, Hwang was photographed collapsed in a hospital bed, unshaven, suffering from depression and exhaustion. The incident was still considered to be one of the biggest scientific frauds in history.

And yet, somehow, Sooam Biotech Research Foundation had survived, crawling out of the hole Hwang had dug, building on the cloning science he had developed. Hwang himself was still at the helm of the company — reclusive, refusing to give interviews, silently building his business in the hopes of reclaiming his standing in the scientific community. To Minh, Sooam was at the forefront of commercial cloning, pushing the science beyond the laboratory on a daily basis. Dolly the sheep was an experiment; Sooam's cloned canines were a real-world, profitable, and in Minh's mind, noble application.

It was those efforts — a formerly lauded scientist, at the forefront of a new technology, trying to rebuild his reputation — that had led Minh to where he was now. Down on his hands and knees behind sandbags against a floor of solid ice, he was waiting for explosives to reveal something so big that it would erase the biggest scandal in scientific history.

"One."

Minh shut his eyes against the searing flash of white light. Then the sound hit him — a vicious crack, like that of a huge leather whip inches from his ears. It nearly rolled him over. Steadying himself, he rose above the sandbags.

The demolition tech was already pushing past the small piles of crumbled ice and rock that had been blown out of the precise hole, using gloved hands to clear the last few mounds of debris. Then the man stood back, giving Minh a clear view of what the blast had revealed.

Even from across the cave, he could see the thick clump of red hair. His heart raced at the sight. The images he'd seen on the ultrasound had been confirmed. The sample was likely to be well preserved, judging from the state of the prehistoric cave lion cubs that he collected. This natural, frozen time machine of a cave was the best environment for finding what they needed.

Most geneticists did not believe that any prehistoric cells found in the Arctic could survive the tens of thousands of years of radiation that had followed their entombment. The DNA in the cells in the specimen in front of Minh could have deteriorated too much to be useful in any cloning experiment. But Hwang and others scientists at the foundation believed that there was still a way to clone an extinct animal:

They didn't need an intact or undamaged cell, just a single intact nucleus, which was more likely to be found in a well-preserved specimen.

Minh took a step forward from behind the sandbags, his legs trembling as he approached the red mound of fur. He could see that the specimen was almost entirely intact. A calf, from the size of it, curled in a fetal position, thick legs crooked beneath its body, its head still partially covered in ice.

Cloning an entire Woolly Mammoth from a single nucleus would be an enormous achievement. And if they could somehow do it before anyone else — before the Americans, who, like everyone else, were now fully armed with the power of CRISPR — the foundation wouldn't just be resurrecting one of the world's most impressive creatures, Hwang and his scientists at the foundation would be redeeming themselves, reviving their own reputations, right along with it.

Chapter Twenty-Two

Late Spring 2013

TWO HUNDRED FEET ABOVE THE PLACID WATER OF A GLACIER-FED LAKE IN BRITISH COLUMBIA, PACIFIC NORTHWEST.

A DH-6 de Havilland Otter seaplane banked left over a liquid glade of pure azure, barely two hundred yards from a jutting wooden dock that was its destination. The steel propellers purred against a fine spray kicked up by the twin pontoons' closeness to the surface of the lake. Inside the lavishly refitted cabin of the plane, Luhan held tight to the cushioned seat beneath her. It was only twelve hours ago that she'd left her apartment in Boston. Here she was, looking out the window at a scene ripped from a postcard, surrounded by leather furniture, cherrywood paneling, and even a crystal and glass bar that would not have been out of place in the presidential suite of a Four Seasons hotel. The guy sitting next to her was some sort of nuclear physicist, the two young men behind her were Silicon Valley millionaires, and the private resort she was heading toward would be chock-full of more of the same.

As the plane hit a pocket of air and jerked up, then down, her fingers whitened as she clenched her seat. She wasn't afraid of flying, and she wasn't afraid of water, although putting the two things together seemed a little mad. But then again, nothing about this trip could be described as normal.

When Dr. Church had offered her the opportunity to represent the Church Lab at a prestigious and exclusive annual private gathering of scientists, businessmen, and forward thinkers from a multitude of tech-related industries, Luhan had jumped at the chance.

The first leg of this junket, the flight from Boston to San Francisco, had taken place on a luxury private jet. Not from Logan, but from a gated private airfield thirty minutes north of the city. No TSA agents, no baggage check — just a polite woman glancing at her passport as she boarded a silver, bullet-shaped airplane, its sleek interior appointed with wood cabinetry.

Then the jet had taken off, rising almost

vertically over the runway, the acceleration pressing her back into her soft lounge chair so hard she nearly spilled her glass of champagne. There were seat-belt lights and a stewardess on board, but nobody made an announcement about tray tables or turning off electronic devices. Her cell phone had worked the entire flight, and she'd sent so many texts to Bobby and the rest of the team, they'd probably thought she'd gone insane.

A few hours later, she'd landed, after a nearly vertical descent. By the time she boarded the Otter she felt as if she were living someone else's life. Biologists didn't travel in leather-lined seaplanes to resorts on glacial lakes in British Columbia.

She was considering asking the physicist what he thought, when her ruminations were interrupted by a loud splash. She was jolted forward, as plumes of water spouted outside her window. Then the plane slowed until it was bobbing gently on the lake, still a ways from the dock. The door to the cockpit swung open and the uniformed copilot/steward for the short trip smiled at her and the rest of the passengers.

"Welcome to Canada." He smiled amiably. "If you could please keep your seat belts fastened as we paddle toward the dock, that would be much appreciated."

Looking out the window, Luhan could see that a caravan of SUVs was waiting, engines running.

Nine the next morning, and Luhan had already been up for more than four hours, most of that time spent outside in the woods, breathing air that seemed supernaturally clean and pleasantly overoxygenated. She'd hiked through redwoods and up cliff faces; she'd traversed part of the lake on an elegant little sailboat; and she'd had Italian coffee and expensive French pastries brought to her as she sat on a makeshift viewing platform, watching through a pair of high-powered binoculars as a bear caught fish in a nearby stream.

She had been so inspired by everything she'd seen that she'd almost forgotten about the frustrations back at the lab, the wall they'd run into involving elephant stem cells. The failure was doubly difficult to face, after all the progress she'd made on the other components of their work. The Mammoth sequence they'd gotten from the Reich lab had been accurate enough for their uses, and she'd had no problem finding matches for the four major traits they had been searching for — ears, subcutaneous fat, hemoglobin, and hair. She'd also found many of the other traits they hoped to implant in their elephant cells — mostly related to hair length and growth, and cold-weather survival. The CRISPR technique had proved suited to the job, and Quinn had demonstrated that the synthetic genes they'd created could indeed be cut into the DNA in the elephant cells. But since they didn't have usable stem cells, they couldn't go any further than that. Without generatable stem cells, they could implant DNA from a Woolly Mammoth into an elephant, but it wouldn't generate Woolly Mammoth traits; they needed stem cells in order to make an elephant a Woolly Mammoth.

Still, seeing the bear in its natural habitat was quite compelling, even though a bear wasn't nearly as compelling to her as a dragon.

When she entered the Western-themed restaurant at the five-starred resort, her thoughts were back on the Woolly Mammoth. If they somehow figured out how to get past their lack of stem cells, somehow did manage to generate the Mammoth traits from the inserted genes, what then? At the first meeting of the Revivalists, that was as far as they had needed to go, but to complete the task Church had set for them, they would have to think even farther along the timeline. They would have to place those stem cells into a fertilized egg, then place the egg into an elephant's womb. The Woolly Mammoth Revival team could spend years using CRISPR to put genes into cells, but that alone wouldn't provide them with a baby Woolly Mammoth.

Between now and Church's end goal of Pleistocene Park, on the other side of the world, they were going to need a fertilized egg and a viable womb. Then they would need a place to take care of a pregnant elephant, and if they were lucky, the baby that came next.

Luhan shook her head, to bring her mind back to her lunch meeting. Moving through the restaurant — wagon wheels on the wall and a pair of shotguns hanging from brass hooks near the ceiling — she remembered what Bobby had said about science fiction being real only when you removed the fiction. During the flight to San Francisco, she'd spent some time leafing through articles that had recently come up on her iPad's newsfeed about what her team jokingly called "the competition" — the South Korean company that had just announced its own efforts to clone a Woolly Mammoth from frozen material they'd pulled from the

Arctic ice. Apparently, they had partnered with a Russian university, and perhaps the Russian government, to try to find genetic material that had somehow been preserved well enough for cloning to be possible.

Like her mentor, Luhan doubted that their efforts could succeed. She couldn't imagine they would find any usable DNA in organic material from thousands of years ago. Dr. Hwang, with his scandalous past, was motivated to make such a grand announcement. But no matter how good a Mammoth looked coming out of the ice, its cells and DNA were likely too degraded to be usable.

Then again, as Church had told his lab over and over, nothing in science was to be written off — nothing was impossible. Motivation — whether it was reputation, profit, ego, or altruism — could spur awe-inspiring innovation.

Luhan reached a long table that dominated the middle of the restaurant and took the one empty seat. A youthful man was sitting next to her: short brown hair, deep-set blue eyes, wearing a suit over a white button-down shirt, open at the collar. Over the next several minutes, half a dozen people came up to him to shake his hand and try to engage him. It was obvious he was someone important. But it wasn't until he introduced himself that Luhan realized she was sitting next to one of the richest men in the world, Peter Thiel.

Peter Thiel's Founders Fund had organized this annual meeting at the Canadian's resort, and the enigmatic billionaire had been a financial presence, both philanthropic and for profit, in genetics and medical science for quite some time. The gathered entrepreneurs and scientists represented just a handful of those whose work Thiel found intriguing.

Thiel had first made his billions in PayPal and Facebook, then branched out, following his passions deep into scientific corners most investors didn't even know about. He'd invested heavily in nuclear research — specifically, experimental fusion — and other forms of "clean" power; he'd also put money into funding efforts to develop artificial intelligence in a controlled, responsible way; and he supported research exploring the bridge between the technological and the biological, commonly known as "the singularity" — the moment when computer technology advanced to the point where it would be possible to "download" humans onto a hard drive, achieving a sort of immortality. He'd put money into politics and social projects — funding sometimes controversial efforts, political candidates, and movements that aligned with his conservative libertarian views. He'd put up a scholarship that encouraged brilliant entrepreneurs under twenty to drop out of school and pursue their dreams, believing that school wasn't the answer for everyone.

But perhaps his most pressing passion, and the one that had put him into Luhan's sphere of expertise, was life extension. Reverse aging, immortality, whatever label the science came under, Thiel was interested, and willing to put some of his billions behind it.

Luhan had never met a billionaire before, and from what she had read about Thiel, she had expected him to be arrogant and intimidating. But she quickly found him quite humble, even charming, and well versed in genetics and the health sciences. His passionate belief that the marriage of science and investment could create astonishing advances — such as true clean energy and massively extended life spans — was infectious. Before she realized what she was doing, Luhan had launched into a monologue about her own philosophies of scientific engineering, colored by the work she and the Revivalists were doing.

"I think we are often stupid in the face of nature," she said, knowing that her English competency was still trailing behind her ideas. "For instance, we didn't invent CRISPR, nature invented CRISPR — it was a bacteria's natural way to defend against viruses. That's all CRISPR is, a natural mechanism of defense that we've borrowed and turned into a tool for genetic engineering. Before CRISPR, our own efforts to edit DNA were so cumbersome, so time-consuming. Then nature gave us an elegant solution."

A line of people had formed behind Thiel, trying to edge into the man's orbit, but he didn't cut her off, so she continued.

"You want to know how we can use science to live forever? The key is, we don't need to start from scratch. We start with nature."

In her mind, she was back in the Church Lab, staring at a Petri dish containing cells from an elephant's placenta.

"Elephants don't get cancer. Why? Well, we don't really know, but it probably has to do with something deep within an elephant's genetic code. There's something in its DNA that keeps the cells, even in such large numbers, such a large biomass, from generating mistakes. Right now, in the project we are working on, we are running into a wall because of it — that same mechanism probably makes it hard for us to generate elephant stem cells, which we need. But one day we might find whatever it is that keeps those mistakes at bay. We might find the natural secret that keeps an elephant from getting cancer, and then we will take that secret and apply it to our own cells."

Suddenly self-conscious, she felt her cheeks grow warm, and it wasn't only because Thiel was listening to her so intently or because the rest of the guests at their table had gone quiet as she continued to speak.

"The same theory will apply to reverse aging. We won't invent the secret — we'll borrow it from nature."

As Thiel pressed her for more information about the project that had led her to her knowledge of elephants, she realized that she was on the verge of something. An answer to their stem cell problem was sorting itself out in her thoughts. What she had told Thiel didn't just apply to his own passions, his goal of infinite life extension. It applied to the Revivalists as well.

The answer they were looking for was already there, in nature, in the elephant cells themselves.

After she had filled in Thiel on more of the science of the Woolly Mammoth project, he congratulated her and turned his attention to another guest. In May 2015, he had already given one hundred thousand dollars to help fund the Mammoth Revival project. Up to the point at which Thiel had stepped in, Church had been funding much of the project from his discretionary monies, along with funding from Revive & Restore. Thiel's deep pockets, and their aligned interests, could help them get to the next stage.

Luhan felt an excitement growing inside her that had nothing to do with money. Without money, much of science was impossible. But money alone wouldn't allow men like Peter Thiel to live forever. Nor would money alone bring the Woolly Mammoth back to life.

Speaking to Thiel, Luhan had realized that she had solved her own problem. And the answer wasn't money. It was nature.

Chapter Twenty-Three

Late Spring 2013

77 AVENUE LOUIS PASTEUR.

At ten minutes past three in the morning, Bobby was still shaking sleep out of his eyes when Luhan burst into the lab, as usual moving like some sort of possessed sprite. No hello, no small talk about the trip to British Columbia on the private jet or the long flight home, no acknowledgment that three in the morning was a hell of a time to call him, to drag him out of bed, to demand that he meet her in the lab. No acknowledgment that Bobby was even in the room. Instead, she headed straight for the back of the lab, to the steel shelves that were jammed between hooded work cabinets, and began leafing furiously through a stack of old science journals and publications, her fingers plucking through the covers as if she were fretting a guitar.

"So how was the in-flight movie?" Bobby

tried, but Luhan wasn't in the mood for small talk.

"Where's Justin? Margo?"

Bobby shrugged and said, "They're probably smart enough to keep their phones on silent after midnight. I'm unlucky enough to have a wife who works a night shift in the ER once a week. Which means nothing in my life is ever on silent mode."

"Doesn't matter," Luhan said, "we'll get them up to speed in the morning."

"Technically, it is morning."

"Bobby, what is aging? Why do we age?"

She was still bent over the journals, searching. Bobby stared at her, wondering where that non sequitur had come from. Then again, with Luhan, nothing was ever really a non sequitur. She could continue a line of thought, or a conversation, picking up after long breaks — hours, days, weeks — no matter how many distractions occurred in between.

"Well, it's a big question. But in part, we age because cell replication causes the breakdown of certain physical properties that cells need to survive."

Bobby was simplifying things to an extreme, but he was guessing that Luhan was thinking of the big picture. There were actually many definitions of aging, and many reasons that people and animals got old. Bobby had spent a lot of time thinking about the processes that caused the mutations and cell death that, eventually, led to a creature's death.

"But is it natural? Is aging a natural process?"

Bobby shrugged.

"I look at it like cancer, or any other disease. It happens in nature, but it's really a breakdown in the natural process of cell division. I believe aging is something we can cure, like any other disease. So in that sense, no, it's not natural."

Luhan nodded. She had expected his response, since she and Bobby had talked about this before. Then her back straightened suddenly and she pulled a journal out of the stack.

"Our elephant cells — we can't use them because they die, just like aging cells. That's why we need stem cells. We need cells that can divide, indefinitely. Like nature intended," she said.

Bobby added, "We also need stem cells because they can turn into all the other cells in the body —"

Luhan cut him off.

"Yes. One problem at a time," she said. "The first thing we need to do is stop our cells from dying. We need to cure them, the same way we're going to eventually cure the disease of aging."

"You want to immortalize our elephant cells," he said.

Making cells "immortal" was a process widely used in biochemistry, in order to better study cell lines. Basically, cells could be immortalized by adding viruses that contained DNA that countered some of the deteriorating processes involved in cell death. The bigger problem — the breakdown of telomeres that protected the chromosomes — could be addressed by adding a stretch of DNA encoding a protein that could help elongate the telomeres, making them last longer.

"It's possible," Bobby continued. "We'd have to remove the additional gene right before we finished with our implantations, or the cell wouldn't express correctly or could develop tumors. But it still doesn't solve our second problem. Maybe we can show our work — get our genes to express themselves as traits, like red hair and working hemoglobin — but without stem cells, we still can't do anything with them."

Adding telomeres — immortalizing cells — would not create a fountain of youth. It would not provide a system-wide immortality. The cells might divide and replicate forever, but they would be in a single cell line. A skin cell that kept going. A hair follicle that never stopped growing. Also, the act of immortalizing the cell affected its biology, introduced opportunities for mutations and contamination.

"That's why we have to take it one step further," Luhan said, holding the journal she had just retrieved in front of her to show Bobby the cover.

Bobby recognized it: *Cell,* from November 2007. He immediately knew which article Luhan had remembered that caused her to drag him out of bed:

"Induction of Pluripotent Stem Cells from Adult Human Fibroblasts by Defined Factors."

Bobby had no problem remembering the title, no matter how complex it sounded. The work had won biologist Shinya Yamanaka of Kyoto University the Nobel Prize in 2012, just six months ago.

"Beginning back in 2005," Luhan said, "Yamanaka and his team began experimenting with skin cells, infecting them with RNA viruses that carried very specific strands of DNA."

"Genetic engineering, pre-CRISPR," Bobby said, as Luhan flipped open to the relevant pages.

"Correct. And slowly, he made an incredible discovery. With the right combination of genes, it was possible to turn a skin cell into an iPSC."

"An induced pluripotent stem cell."

Bobby felt a jolt as the full import of the article hit him and drove the last remnants of sleep from his body.

"He made stem cells," he said.

"And that's exactly what we're going to do," Luhan said. "Yamanaka narrowed it down to just four genes that needed to be added to the skin cells to make stem cells. Through trial and error, he proved that just the right combination of DNA implanted into these cells could change their nature, permanently."

"The four genes — the Yamanaka factors. Oct4, Sox2, cMyc, and Klf4."

"The process is really quite simple," Luhan continued, looking through the pages. "We isolate our elephant cells. We immortalize them. We insert the four factors using CRISPR. We culture the cells. And then we draw out the ones that become stem cells, and start to generate colonies."

Bobby whistled. It didn't sound simple, but it was a process all the Revivalists had grown adept at performing.

"Then we add our synthetic Woolly genes, and we're good to go," he said.

Good to go. It was a funny way to look at it, because all they'd have at that point was a stem cell with Woolly Mammoth properties. They wouldn't have a Woolly Mammoth.

Bobby was reminded of a conversation he'd had recently; he'd been telling someone about the Mammoth project, and she'd asked, "Well, where is your Woolly Mammoth going to live?"

"A Petri dish," he'd answered.

Luhan dug through the journal, looking deeper into Yamanaka's work.

Of course, if this was going to work, they were going to need more.

"We'll need plenty of cells," Bobby said. "And eventually, if this actually succeeds, we're going to need a place to put them, beyond a Petri dish."

Luhan finally looked up.

"We're going to need more elephants," she said.

Chapter Twenty-Four

Summer 2013

POLK CITY, CENTRAL FLORIDA, FORTY MILES WEST OF ORLANDO.

At three in the afternoon, the sun was high above a two-hundred-acre complex of single-story, barracklike buildings, large feeding troughs, mud-floored exercise pavilions, outdoor obstacle courses, and shaded pagodas.

"Okay," Church said to Stewart Brand, through gritted teeth. "I think I'm ready."

Brand was standing behind him, a frightening grin on his angular face. As when they'd met at Brand's bird sanctuary in Petaluma, the Whole Earth icon was wearing his signature safari clothes: tan shirt covered in pockets, scuffed pants, a wide-brimmed hat, and a pair of hunting knives resting in scabbards attached to his belt. But the device in his outstretched hands was not a knife. It was a cattle prod, a couple of feet long, mostly plastic, with two terrifying metal prongs sticking out one end. At the moment, the prongs were only a few inches from Church's upper left thigh.

"You sure about this?" Brand asked.

Church's hands were balled into fists, his eyes tightly shut, as he hunched forward, offering up as good a target as he could muster. He was wearing jeans and a T-shirt, but he didn't think the clothes would make much of a difference

"Anything for science, right?" Church said.

"Better you than me," Brand said, winking.

He touched the baton to Church's jeans. The pain was instantaneous, shooting down Church's leg and up his back, electric tendrils reaching all the way to his beard. He gasped, but didn't cry out. Then, just as suddenly, the pain disappeared.

He straightened, then ran a hand through his frizzed-up hair.

"Not as bad as I thought," he said. "You want a turn?"

Brand glanced at Phelan, who was talking to a man with a clipboard over by a high metal gate. She rolled her eyes at them, and Brand turned back to Church.

"I think I'll take your word for it."

Church laughed, rubbing his thigh. Getting zapped by a cattle prod wasn't nearly the wildest thing he'd ever done just to prove a point. Long ago, he'd spent days walking through his lab wearing blinders, to show his lab mates what it meant to have tunnel vision. When he was becoming vegan in 1973, he'd subsisted for quite a while on an entirely synthetic diet that he'd whipped up in his lab, just to see if he could.

Now that he was going to be working with elephants, it made perfect sense for him to try out one of the controversial electric prods that many elephant handlers used to train and control the majestic beasts. And although the shock had been painful, Church didn't think it constituted animal cruelty. He knew that some elephant trainers could — and did — turn up the voltage well beyond what he'd just gone through, but then again, he wasn't an elephant. But he did feel that, at the moment, he was standing in pachyderm heaven.

When Ringling Brothers and Barnum & Bailey Circus had first contacted Church a few weeks earlier to visit their Center for Elephant Conservation just outside Orlando, Church had been both surprised and thrilled. Apparently, the people behind the "Greatest Show on Earth" had read one of the many interviews he had done on the Revival project, and had been immediately intrigued.

The timing couldn't have been better. Luhan, Bobby, and the rest of the team had immortalized cells and synthesized stem cells, which meant they were getting closer to the time when they'd need a real partner with access to many Asian elephants. Meanwhile, the circus had recently entered a new phase in its relationship with its central attraction. Responding to a growing popular movement toward eliminating the use of elephants for commercial and entertainment purposes, they had recently decided to retire all of their performing elephants.

Church had mixed feelings about the circus's decision; as a kid, he'd grown up among snakes and starfish and he'd seen larger animals only from visits to the traveling circus. He'd always believed that people became interested in conservation and preservation because they, too, had seen these great beasts in their own home towns. But public opinion about the circus had changed, and there was a widespread belief that using large wild animals in circus acts was cruel, even if the professionals who worked with them were, on the whole, kind and devoted to their welfare. But Ringling

Brothers' retirement of the elephants, the circus's loss, could be the Revival team's gain.

Brand and Phelan had asked to join Church on the trip to Florida. As Brand described it, he wanted to touch the elephants, see what it was like to be with them, and think about them one day becoming Woolly Mammoths. So the three of them had traveled to the two-hundred-acre compound, home to nearly forty retired Asian elephants, the largest herd on the entire continent.

The Conservation Center was actually small in acreage, compared with other refuges. Church knew of at least two that were spread over thousands of acres. But Asian elephants didn't need the same amount of space as their African brethren. They were indigenous to the Asian jungles, not the African plains, and used to smaller quarters.

Church, Brand, and Phelan felt the poignancy of seeing such majestic beasts in captivity. It wasn't like viewing horses in a pasture; these were large, wild animals that had to be kept safe, and their care involved a lot of hardware. The handlers employed by the circus had been working with the elephants for decades, and had loving relationships with them. Each elephant was individual, unique. But controlling a large population — really, a super-herd — took engineering and science.

Church, Brand, and Phelan spent hours strolling through the various barriers that had been erected to control the animals' movements, from the many fences and enclosures to specially designed blockades with openings big enough so that people could run through them if they were threatened, but too small to allow a charging elephant to follow. They took stock of the cameras, locked doors, and feeding zones, and last, the metal hooks and prods the handlers resorted to when the elephants were unwilling to correct or redirect natural behavior that could be extremely dangerous in an enclosed environment.

At first glance, the batons hadn't struck Church as particularly cruel; similar devices were used on farms all over the world. So he'd asked if he could try one out himself. The handlers had assumed he'd meant on an elephant, but Brand had understood his intention right away, and had happily volunteered to apply the charge.

Church also had the chance to check out the circus trains in which the elephants had traveled, when they were still performing.

Though small, the cars seemed comfortable, if not luxurious. Every effort had been made to make the elephants as happy as possible.

The Conservation Center included a full-fledged habitat outdoors, with sloping hills and wide-open ranges, a well-designed obstacle course, even a massage and stretching center for the herbivores' physical and mental well-being.

The handlers had told Church that the center spent more than seventy thousand dollars a year taking care of each elephant. Brand wasn't surprised at the attention to detail. Everyone he'd ever met who'd worked with elephants had developed an enduring love of the species, especially the Asian breed, who were smart, agreeable, even friendly. The way one vet at the center had described it, when you surprised an African elephant, it either charged or fled. When you surprised an Asian elephant, it responded with curiosity.

Of course, even the gentlest of beasts was still formidable. After the experiment with the electric prod, Church and Brand followed one of the handlers into a gated area just past one of the feeding zones. A large male elephant was off to the left, kept behind a row of massive columns in the ground that created a five-foot-tall barrier. The columns had been designed to separate male elephants from the rest of the herd when they were in musth, or heat — rutting season. When a male Asian elephant is crazed with lust, it can break through all but the strongest enclosures to get to a female, and it will trample anything and anyone in its way.

On the other hand, when a male elephant is in heat, it is a great time to get samples of semen for the Conservation Center's genetic database, which it had been adding to for more than a decade. While Church stood by the columns that were holding back the bull elephant, Brand wandered over to an odd-looking device that reached almost to his shoulders. As he was running his hands over its curved plastic, he realized it was designed to look like a female elephant's rear end.

"Uh, Stewart," Brand heard Church mumble, but he was too busy trying to figure out the device.

"Hold on a second. This is quite remarkable."

"Stewart."

"Ah," Brand continued. "I think this is an insemination machine. The male elephant goes in here . . ."

"Stewart!"

Brand turned around and saw that the bull elephant had gone completely ballistic — his eyes rolled back, his feet pounding at the metal pins. He wanted to stomp Brand into dust.

"I think you're messing with his girlfriend," Church said.

Brand jumped back from the device, his face turning red.

"I guess jealousy is universal. When the time comes, remind me never to get between a Woolly Mammoth and his mate."

Church laughed, as the elephant eyed them angrily. He apologized for his friend, and then both he and Brand followed the handler out of the insemination pen.

A few hours later, Church, Brand, and Phelan were sitting in a conference room located in one of the low-slung buildings at the edge of the compound, kitty-corner to a pair of genetic labs that would not have been out of place at Harvard Med School. Church had been impressed by the circus's research facilities, which weren't just some sort of glorified zoo, but part of a scientific endeavor to help the endangered Asian elephant, both in the wild and in captivity.

The genetic database the center had created was proving helpful in understanding and increasing fertility in the dwindling species, as well as learning the specific health risks facing the giant beasts. Even though the elephants were retired, they now had much more to offer humankind than entertainment. Their cells would aid in the search for cures for cancer and for extending life.

Although elephants rarely got cancer, more than half of humans developed the disease at some point in their lives. As a scientist working on reversing aging, Church thought, *What good would it do to extend life spans, if eventually everyone got cancer?*

Church was convinced that elephants' immunity to cancer was genetic. Something was hidden in their DNA that gave them resistance, that allowed their cells to divide without the sort of mutations that gave rise to tumors in humans.

Given enough time, and enough access to elephant cells, Church was certain that biologists would eventually discover that secret.

Talking about meeting one of the young elephants, a calf named Mike, though, Brand was emotional.

"He's pretty hairy, so much more hair than I'd expected. And so beautiful, already big and muscular, but so gentle. Just like the grown elephants, the first thing he did was sniff my shoes with his trunk, to learn where I'd been."

As a conservationist, Brand found walking among the endangered elephants to be a peak experience. Approaching Mike, he'd first been looking at the animal from a purely scientific perspective; sizing him up physically and figuring out where he stood in the hierarchy of the park's herd. But then he'd come face-to-face with the calf, who had playfully reached out to him with his trunk, making an instant emotional connection.

Brand knew that his experience wasn't unique at all — it was well documented that elephants lived extraordinarily complex and emotional lives, both as individuals and as part of herds with an established social order. He had recently read the story of a mother elephant who, during her herd's annual migration year after year, had returned to the exact spot where her baby had died; there was no question that the mother was mourning her loss. Videographers had captured many similar incidents of elephants mourning and even burying the remains of herd members, and elephants protecting and assisting injured or sick relatives. Elephants also developed real, emotional bonds with humans, sometimes lasting decades.

It wasn't just their emotional capacity that struck Brand as special: Elephants were extremely intelligent. They could use rudimentary tools — sticks, branches, rocks — and could even learn vocally, imitating sounds such as whistles and horns. And elephants were playful, and curious; Brand had spent nearly ten minutes joyfully watching Mike figure out, and then drink from, a garden hose. He'd also seen pictures of other elephants encountering snow for the first time, using their trunks to roll huge snowballs to throw at each other. Brand was not anthropomorphizing — he could see the intelligence and emotion in Mike's eyes.

Then one of the center's senior researchers brought him back down to earth.

"Unfortunately, a year from now Mike will most likely be dead."

The researcher went on to explain that the Asian elephant was facing a species-wide catastrophe; a vicious herpes virus had spread throughout the entire population, and unlike the human version of the disease, elephant herpes was often fatal. In fact, more than a quarter of all young Asian elephants would die from herpes, more than were killed by ivory hunters and loss of environment combined.

Hearing the news, Church had immediately caught Brand's eye.

What good would it be to engineer a baby Woolly Mammoth if it immediately died of herpes?

"We need to do something about this," Phelan started.

Before she could continue, Church cleared his throat.

"We can solve this."

His mind was already on overdrive. Nobody had ever been able to culture the virus that caused herpes in a lab, which was the reason nobody had yet come up with a vaccine. But every scientific problem had a solution, a workaround. His Revival team was working around a lack of stem cells by immortalizing regular cells and turning them into synthetic stem cells. He could work around the inability to culture the herpes virus — by writing it himself.

"We'll study fragments of the virus, then we'll reconstruct its genome. Write it in the lab, then grow ourselves a vaccine."

The researcher stared at him. It was classic George Church.

"If we're going to ask an Asian elephant to help us bring back the Woolly Mammoth," Church said with a shrug, "it would be nice if we could do something for them, too."

Quid pro quo.

Just like that, Church had laid out a pathway to the cure for elephant herpes — and the Revivalists had their elephants.

Chapter Twenty-Five

Winter 2014

SEOUL, SOUTH KOREA, SOOAM BIOTECH RESEARCH FOUNDATION.

A marble and glass complex at the edge of a perfectly manicured lawn, protected by a gate and a pair of uniformed guards. Jy Minh shivered as he moved through the surgical theater, his blue scrubs swishing against the elastic ties of his sterile rubber slippers. His mask and surgical cap felt tight — perhaps he had been spending too much time in the field, garbed in yak fur and down coats the size of sleeping bags, to remember what it was like to work in a biology lab. Then again, the central cloning floor of the Sooam Biotech Research Foundation was not a common biology lab.

He shivered again as he squeezed between the operating tables. The air was almost as cold inside as the winter weather outside; the windowless interior of the sterile room had to be kept at an optimum low temperature for the specimens — not necessarily for the surgeons. But his body wasn't reacting to the breeze that the high-tech ventilation system was blowing. No matter how many times he'd been through that operating room, no matter how deep his training, no matter how many animals he'd cut into over the years, dead and alive, he'd never gotten used to the setting.

Three tables were set up at four-foot intervals across the large rectangular theater. Technicians dressed the same as Minh — blue scrubs, rubber boots, masks, and caps — were gathered around each table, checking and rechecking intubation tubes, blood lines, anesthesia gauges. The surgeons moved between them, shifting from table to table, making sure the subjects were properly prepped for their procedures.

As Minh passed the closest table, he could see the brown fur and limp tail of the first animal's hindquarters. Most of the body was covered by a tarp, with surgical windows placed at exactly the precise points where the surgeon would cut; the animal's head hung over the other side of the table, the jaws pried slightly open to allow the insertion of the intubation tube.

Minh had been in many hospitals before,

had witnessed surgical procedures many times. He'd even watched his own wife's Caesarian, years earlier; he'd sat by her head on a little stool, holding her hand, while an obstetrician cut through her abdomen to the womb to rescue their second child from a troubled labor.

But somehow, this procedure was different. This seemed — unnatural.

He passed the second table, where the surgeon had already begun making a tiny incision through the window in the tarp. For the surgeon, this was an easy, routine procedure that he had performed hundreds of times before. In a human, it would have been a similar operation with similar equipment, a tube and an ultrasound probe, a suctioning needle and a steady hand.

The third operating table was slightly different. There, the incision had already been made, the catheter inserted, and the solution was ready to be pumped directly into the uterus. Here was the end result of the work begun on the first two tables. Compared to the harvesting, the insertion was much more delicate, and took the trained and steady hand of an experienced professional. Even so, the procedure was successful only one out of every three times. There was a better than even chance that the insertion wouldn't take or that the developing embryo would die before it reached term. Even if the embryo came to term, many of the animals would not grow to a healthy adulthood. In a room downstairs, veterinarians tried their best to nurse the less perfect specimens back to health.

Although the process was not perfect, and Minh couldn't shake the feeling of discomfort that hit him as he moved through the lab, he knew he was watching a medical miracle. The work that took place in between the first two tables and the last was, in his opinion, Nobel Prize–worthy — although, because of the scandal and the claims of fraud surrounding him, his boss, Hwang Woo Suk, the man who had pioneered the practice, would never be eligible for that award.

But in the midst of his questioned work, he'd also achieved something spectacular. In August of the same year as his trial — 2005 — he'd published a paper in *Nature* on the successful cloning of an Afghan hound: a dog named Snuppy, a breakthrough so amazing that *Time* magazine named the dog the year's "Most Amazing Invention."

Although his human cloning claims were false, his cloned dog was real. Hwang, an obscure veterinarian who had previously spent his time working with cattle, had done what nobody else had been able to accomplish: He'd cloned a dog.

Since then, his revived foundation had replicated the feat more than six hundred times. Minh had watched the intricate process many times. The harvested eggs from the dogs in the operating room were placed in a Petri dish. A pipette was used to remove the eggs' DNA, and then the DNA from another dog — usually, from a living skin cell — was inserted, through a process called "somatic cell nuclear transfer." The egg was then stimulated with an electrical charge, forcing it to divide, essentially kick-starting its growth into a functioning embryo. Then the sample was brought back into the operating theater and carefully placed into another dog's healthy uterus.

Life from life, Minh thought, as he moved past the implantation table and headed toward a pair of double doors in the back of the operating theater. Still, he couldn't shake the chills he felt as he thought of those three dogs, lying supine, tongues hanging to the side around the breathing tubes, as the surgeons did their work. He chided himself for his own backward thinking; of course, futuristic science always seemed unnatural — until it became routine. Hwang wasn't simply providing a service to grieving, wealthy dog owners; he was also pushing forward the science of cloning, giving the world a glimpse of what this powerful new technology could provide.

His cloned dogs could help in the fight against diseases like Alzheimer's, cancer, and diabetes by providing identical physiologies — perfect control specimens — for drug studies and the like. Hwang had also gifted to South Korean police departments "sniffer" dogs that had been cloned from the skin cells of talented working animals that had been trained to sniff out cadavers, bombs, and drugs. All dogs have a keen sense of smell, but certain breeds of dog are more responsive to training and fieldwork. So by harvesting genetic material from dogs with these superior natural abilities, Hwang could provide the police departments with animals that had literally been born for their jobs.

Hwang was planning to send the specially bred dogs to the foundation's partners working on a project in Russia, as well — a sort of altruistic precursor to the more complicated cloning subjects they hoped to deliver in the future, although many would argue that nothing Hwang or Sooam did was truly altruistic. The human cloning scandals would not be overcome by the foundation's work with cloned dogs.

Minh pushed through the double doors, then headed down a long hallway leading deeper into the research complex. Opening a locked, vaultlike door, he entered an austere office lined with bookshelves, microscope cabinets, and computers.

For the past year, this had been Minh's center of operations when he was not out in the field. Although he was not secretive by nature, Minh and other researchers had learned from Hwang's past of the danger of presenting findings to the world that hadn't been confirmed.

Minh reached his desk in the center of the room and dropped heavily into a leather-backed seat. Immediately, he turned his attention to a manila folder containing a stack of pages next to his computer.

Months ago, when he had returned from his most recent trip to Siberia, the baby Woolly Mammoth specimen he had found was taken to a specially designed tank in the research facility. He had overseen the transfer and storage process himself, making sure the air temperature, humidity, and sterilization were all precisely maintained. He had believed, when he'd first seen the carcass pulled up from the ice, that the specimen was very close to what they had been looking for — perhaps as good as they would ever find.

But if the papers now in his hands were correct, he had been dead wrong.

According to the documents, in May 2013, a Russian expedition led by the foundation's partners at Northeastern Federal University in Yakutsk had pulled another Woolly Mammoth out of the ice on a remote island above the Arctic Circle. Although that specimen had been spectacular enough — a female, sixty years old at the time of her death more than fifteen thousand years ago — when they hoisted the Mammoth out of the ice, they'd made an even more significant find.

When the animal had died, her lower part had settled into water, which had quickly frozen. Although the upper part of the beast had been exposed to the elements, and a large portion of its meat had been eaten away by predators, the lower segment of her body had been preserved in a manner the Russian scientists had never seen before.

Minh read, and reread, the words of the lead Russian, Semyon Grigoryev:

"When we broke through the ice beneath her stomach," Grigoryev had exclaimed, "the blood flowed out from there. It was very dark."

To Minh, the comment was shocking. A carcass so well preserved, after fifteen thousand years in the ice, that it still contained liquid blood? Reading deeper, the Russians described finding intact muscle tissue that was "red, the color of fresh meat."

Minh's find in the ice cave at Muus Khaya had been stunning in its completeness, and although he hadn't yet finished extracting cells from the specimen, he had hope that he would find some informative, if not usable, cell tissue. But red muscle tissue? Blood?

Actual liquid blood?

It didn't seem possible. Minh knew, better than most, that scientists under pressure to make discoveries could exaggerate findings. They could also be so excited by a discovery that they misinterpreted what they saw. Hell, the entire building around him — the kennels down below filled with cloned puppies being weaned by surrogate mothers, the operating tables he'd just walked past, the labs filled with frozen specimens from the Arctic — all of it had been built after a scientific exaggeration.

But if the Russians had found a Mammoth with actual blood — and intact DNA

— it could be the breakthrough Minh and his colleagues had been waiting for.

Minh put the manila folder back onto his desk and pulled his cell phone out of his pocket. He had been back in Seoul for only a few months, but he had no choice — he needed to book a flight to Russia at once. He needed to see this new discovery for himself.

Many geneticists thought that the goal of cloning a Woolly Mammoth, which Hwang and his research center had set for themselves, was impossible. Many questioned their motives, believing that Hwang would overreach or exaggerate the significance of his work again in order to revive his destroyed reputation. Other labs competing to resurrect the Mammoth had chosen the synthetic route over cloning because they believed that DNA could not be revived after millennia in the ice.

The blood flowed out from there. It was very dark.

According to the papers on his desk, the Russians had hidden their sample in some sort of secret vault, because the lead scientists believed that it was so valuable someone might actually try to steal it. To Minh, that seemed overly dramatic — and a little suspicious. Then again, even in America, scientists could be secretive, even paranoid. People stole ideas from each other all the time.

Certainly, Minh's superiors at Sooam understood the importance of secrecy. Only recently, the foundation's parent "chaebol," or mega-company, had completed the purchase of twenty thousand acres of farmland in Alberta, Canada, that seemed a perfect sister site to the Siberian tundra regions, a secluded and vast tract of geography well suited for a revived Woolly Mammoth herd. When local residents reacted with suspicion and fear that the Korean company was actually planning some sort of mining expedition in the region, the company had released a beautifully obtuse statement claiming the land would be used for "experimental agricultural techniques." Even so, the company's attempts to avoid any details about the purchase had led some locals to wonder if Sooam was building a "Jurassic Park" in northern Canada.

Minh wouldn't know for sure what the Russians had actually found until he saw the specimen for himself. But for the moment, he dared to be optimistic. There was a small chance that a fifteen-thousand-year-old Mammoth had just made the impossible possible.

Chapter Twenty-Six

August 15, 2014

The symposium in the Harvard auditorium appeared to be in full swing as Church reluctantly followed Ting along the edge of the full auditorium, searching for a pair of empty seats. He was surprised to see so many people in the room; there had to be at least fifteen rows of seats set up in front of the podium, and maybe 150 people. The room was dark, most of the light coming from the projection screen that had been pulled down for the presentation, but Church was pretty sure he recognized many of the faces. Geneticists, biologists, some fairly big names. Which really didn't make much sense, considering the topic.

"This many people came to a conference about cat DNA?" he whispered, but Ting waved at him to be quiet. She was determined to find them reasonable seats, and was moving along the side of the room at twice her normal speed. Church, with his long legs, was struggling to keep up.

"But cat genomics?" he whispered again. "Really?"

Sure, once upon a time, the sequencing of a common house cat would have been big enough news to fill a Harvard auditorium. Back in 2007, when the folks at Cold Spring Harbor had sequenced a four-year-old feline named Cinnamon, the effort was another example of the sequencing process that had already yielded workable genomes for humans, chimps, mice, and dogs, among a handful of other animals whose cells had found their way into the hands — and hypodermics — of scientists at gene labs all over the world. But seven years later, Church wasn't aware of anything groundbreaking going on in the cat world that would draw such a crowd for a single speech, let alone a three-day conference. When George had first seen the bulletin advertising the GCAT60, it had hardly registered.

"We've gone from karyotyping and pictures of the chromosome," the speaker was droning on as Ting finally found a pair of chairs in the back row, "to a sequence, to a full phylogeny, to a radiation hybrid map, to a genetic linkage map, to sixty sequenced cat genomes. We now have the cat genome as a template to sequence other species . . ."

Church shifted against the hard plastic of his chair. He could already tell that this session was going to tax his narcolepsy. But people tended to watch him for his reaction, and he didn't want to appear as uninterested as he was. Also, the speaker, Fritz Roth, wasn't some hack biologist. Roth was a professor at the University of Toronto and had earned his Ph.D. in biophysics at Harvard in Church's own lab.

In fact, as Church glanced at the people in the row around him, he recognized other faces — more former Ph.D. students of his, most of whom had gone on to great careers in genetics at other universities.

Maybe there had been some new development with the cat genome that had squeaked past him. Recently, he had been consumed by other research. If the symposium had been about elephants, he would have arrived twenty minutes early to sit in the front row. Since his visit to the Ringling Brothers Conservation Center, he'd put the Revival project front and center.

"We'll go over the selection pressure in the cat genome," Roth said, although his voice was mostly background noise to Church. "And then we can think about the ancient history of domesticated cats . . ."

Church's mind was in the lab with Luhan and Bobby. Quinn had taken a temporary leave from the team — his mother was extremely ill, and was participating in a clinical trial — and Margo had recently left for work in scientific patent legislation. But Luhan and Bobby had been preparing numerous cultures of elephant cells for the implantations, creating immortal cell lines with the necessary properties to become iPSCs (induced pluripotent stem cells). Soon they would be showing Church elephant cells with partial Woolly Mammoth genomes.

Which meant it was time to start thinking farther ahead and to consider more of the ethical dilemmas they were soon going to face. Church had set up a special team to work on the cure for elephant herpes and the team was synthesizing a version of the virus, and was working on a way to deliver its cure into the affected cells. Church had come to the conclusion that building a Woolly Mammoth, in some ways, was secondary to helping the endangered elephants — it wouldn't do any good to create a Mammoth if it harmed the elephant population in any way, essentially trading an endangered animal for an extinct one. The work had to benefit both species.

Which brought him to the biggest ethical dilemma they were going to face: once they'd managed to create a faux stem cell containing Woolly Mammoth traits, they would need to implant it into a fertilized embryo, and somehow get it to term. But using a pregnant female Asian elephant was ethically troubling. No doubt, there would be numerous starts and stops. When cloning an animal, miscarriages and mutant births occurred more often than normal. Even Sooam's dog factory, as Church thought of it, claimed successes only one-third of the time. The Revivalists couldn't continue their project with a rate anywhere near that troubling. It would be much too hard on the elephant mothers.

So Church had begun to think of a solution, one that both thrilled and terrified him.

"Now let's get into some defining traits," Roth was saying, as the screen behind him showed two house cats in different poses, one on its back with a blissful look on its face and closed eyes, and another whose back was arched, tail up, and fur erect, snarling. "Gentle behavior versus aggressive behavior."

Church momentarily lost his train of thought. A new pair of slides appeared, one of a hugely overweight cat and one of a cat so small it fit into the palm of a hand.

"Giant body mass versus a smaller cat."

Church glanced at Ting. This was getting appalling. The silly pictures on the screen had obviously been pulled from the Internet, stills from funny cat videos. Had Roth lost his mind? Then Church noticed that Ting was smiling, and that all the other people sitting around him were looking at him.

"We realized there is a human analogy," Roth continued.

A picture of a cat with a white beard appeared on the screen, and the audience started to laugh.

"We begin to let the cat out of the bag," Roth said.

The cat morphed into a picture of George Church. The letters below the cat — the title of the symposium, GCAT60, began to move, until Church suddenly realized what they really stood for.

GC-AT-60. George Church At 60.

He'd been so caught up in his work, in elephants and futuristic ideas, he had forgotten that it was only a short time before his sixtieth birthday.

No wonder he had recognized many of the faces in the room. As the lights went up, he realized that everyone there was either a former member of his lab or a former mentor. He even saw Sung Hou Kim from Duke near the front of the room, preparing to give his own lecture on the cat genome.

The conference was a joke on him. As applause rang out, champagne bottles were uncorked. Church rose to his feet as his current lab members filed in from an adjacent room where they'd been watching on closed-circuit TV. Church shook his head, amazed that such a large group of people had managed to pull this off without his getting wind of it. He was truly touched.

"Well, I am surprised," he said, as the applause died down. "Used to be a big deal, the sequence of the cat genome."

Then everyone was congratulating him, shaking his hand, reintroducing themselves. It was incredible to see the many tendrils his lab had sent out to so many different universities in so many countries. Much of the future of genetics was embodied by the scientists in this room. Church felt no small sense of pride that his lab had seeded the original work and ventures the assembled scientists represented.

It was close to an hour before he'd worked his way through most of his former charges and found himself in a corner of the room with Luhan, Bobby, and Quinn and Margo, who had joined for the surprise party. Church felt it was time to talk a little shop and tell them all that he had decided, even if Luhan and Bobby would be doing the heavy lifting going forward.

Church got to the point, as direct as always.

"I've gone over it, and there's just no ethically responsible way we can experiment on fertilized embryos inserted in female Asian elephants."

The postdocs seemed surprised. As lab scientists, Luhan and Bobby might have weighed the importance of their work more heavily than their responsibility to the comfort of the individual elephants. As animal lovers, Margo and Quinn understood.

"Where does that leave us?" Luhan asked.

"What's the workaround?" Bobby added.

There was always a workaround.

Church smiled at a passing former Ph.D., who lifted his glass of champagne in congratulations. The man now lived in London, where he was working on a universal flu vaccine that would eventually save millions of lives. Church focused back on his team.

"We're going to make a synthetic uterus," Church said.

Bobby gave a low whistle and said, "An artificial womb."

Bobby and Luhan had been working on a fertility project on the side. Because of Bobby's own interest in having a child, he was well versed in the current state of IVF and reproductive science. Although it was technically possible to grow embryos outside a womb for a significant length of time, no scientist had yet tried to grow one past fourteen days. The simple reason was that it was against the rules. A self-imposed ban on growing human embryos past two weeks had been agreed upon by scientists all over the world, and many countries had passed laws to ensure the guidelines were followed.

The moratorium on research past fourteen days, which originally had been agreed upon back in 1995, though ethical, made studying implantation and gestation exceedingly difficult.

"Is the science there yet?" Luhan asked.

Growing a baby outside a natural womb was possible; preterm babies were being kept alive at twenty-three weeks. But those first twenty-three weeks were the challenge.

"I've looked into the literature," Bobby said. "I haven't found anything online or in any journals."

"That's because it's never been done,"

Church said.

"But we're going to do it," Luhan finished for him.

They would file a grant to make a synthetic uterus. Church knew the grant committees would balk at the idea — some would say that it hadn't been done because it couldn't be done.

Maybe they'd be right — but that had never stopped Church before.

Chapter Twenty-Seven

June 20, 2016
NANTUCKET SOUND.
Three p.m., halfway across from the island to Hyannis, the double-decker Hy Line fast ferry cut at thirty knots through a low chop, white-capped jags spitting spray against the sleek, angled hull.

A seagull was suspended a few yards from the railing of the sun-splashed outer deck, frozen against a backdrop of celestial blue, like a prehistoric creature trapped in glacial ice. Wings spread against the surface drafts, feather skirts tilted into perfect airfoils that maximized lift, eliminated drag, the seagull was held in a complex, motionless balance, exactly matching the Hy Line's speed and direction.

If Ting had climbed up the shoulder-height railing and held out her arm, she could have plucked the bird from the air. Instead, she was content to watch it, until finally a gust of wind from an errant wave rustled through the bird's feathers, shattering the imagined glacial ice, and the seagull dived straight down the side of the ferry, plunging toward the water below.

"We seem to spend an inordinate amount of time leaning over railings," Church said, next to her, as they both watched the bird touch the chop, then rocket right back upward, something green and stringy hanging from its beak.

Ting laughed. The truth was, these moments had become rare. Both she and Church were in peak periods of their careers, which meant many long hours in their respective labs, as well as many days on the road, pushing their work at symposia and to colleagues all over the world. The trip to Nantucket had been a chance to combine their areas of expertise, a coming together of two very different perspectives on a shared biological future.

As on their very first date, Ting knew that she and Church could be looking at the same thing over that railing, seeing something completely different, and yet be completely aligned. He might be thinking about the mechanics of flight, while she was thinking of the genetics. Or she might be thinking about the avian visual system, while he was thinking about the chemistry.

This time, Church wasn't seeing the seagull as a whole, but each cell that made up that bird. His vision pierced deep down into the nuclei, to the twisting strips of DNA that coded for every trait that kept the bird aloft. Microscopic sequences of genes generated the shape and texture of the feathers, coded for the strength and depth of the muscles and tendons in the wings, the sharp angles of its beak, and the shape and color of its eyes. Each of those particular cells, derived from the immortal stem lines, was itself the product of at least 50 million years of inheritance. A genome of a billion base pairs passed down from gull to gull, containing thousands of tiny sequences that made the bird everything that it was.

Change even one of those sequences — cut the tiniest segment of that DNA away using CRISPR and replace it with a synthetic gene of your choosing — and you changed the seagull. Change enough of those genes, and it wasn't even a seagull anymore. Make those changes in its stem cell line, and the manipulation entered the species' line, carried forward generation after generation, for another 50 million years.

"It sounds like the presentation went very well," Church said, resuming the conversation they had started while waiting in line to board the fast ferry, the beginning of their short trip back to Boston. He was referring to a town hall meeting that had taken place on the island two weeks earlier. "According to everyone I spoke to who attended, the audience seemed to understand."

Although they hadn't attended the presentation itself, and were getting a feel for the way it had been received secondhand, they hadn't expected to hear that things had gone half as well as they apparently had. They had just strolled through Nantucket's narrow streets and wharves, passing the quaint, cedar-shingled cottages and shops jumbled densely together in concentric waves spreading out from the wooden latticework of docks, feeling transported back in time 250 years. Despite the crowds of summer tourists clutching ice cream cones from one of the half dozen parlors lining the waterfront, and the parents with strollers valiantly fighting the cobblestones as they shuttled children between the various beaches and high-end resorts that spotted the idyllic, upscale island, Ting could easily picture the place as it once was, an eighteenth- and early-nineteenth-century whaling village originally settled by the English, dominating the whale-oil industry that partially powered the American experiment. Derived from blubber, whale oil was the premier energy source of the time, and was once considered irreplaceable.

"They didn't run Esvelt off the island with pitchforks," Ting said. "Which I think qualifies as a good start."

The historic town had been an odd setting for Kevin Esvelt's presentation. The town hall–style meeting took place at the end of a cobblestone path that whaling captains might have strolled down on their way to the docks. Kevin Esvelt, an assistant professor at the MIT Media Lab, who was previously a postdoc in Church's lab, and the organizer behind the event, had addressed the crowd of about twenty townspeople, civic leaders, and members of the Nantucket board of health on the topic of an innovative new solution to the spread of Lyme disease.

Lyme disease is a particularly nasty bacterial infection that affects around three hundred thousand people a year, mostly in the northeastern United States. Symptoms can start off like a mild flu with fever, joint pain, fatigue, and sometimes a distinctive bull's-eye-shaped rash. It is difficult to treat, and if it isn't caught early, can cause years, or decades, of systemic, chronic health problems. People and dogs contract Lyme from the bite of a certain type of tick: *Ixodes scapularis,* more commonly known as a deer tick.

Nantucket and its sister island, Martha's Vineyard, were at the center of the epidemic. According to the Nantucket board of health, more than 40 percent of the year-round population of Nantucket had been infected with Lyme disease, and during the summer, hundreds of new cases were seen every week.

For an economy that lived on tourism — the population in Nantucket swelled from its base of ten thousand year-round occupants to more than sixty thousand during the summer months — the tick-borne disease was a growing, existential threat. In previous years, drastic steps to fight the disease had been discussed, such as mass spraying for the insect, which was tactically difficult, given the island's dense population mixed in with thick foliage, and a culling of the deer population, an idea met with fierce opposition. People chose to live on Nantucket to be closer to nature, not to destroy it.

But two weeks ago, Esvelt had offered a

third solution, one pulled directly from Church's playbook. Indeed, although Ting and Church had missed the presentation, they had spent part of their weekend on the island getting feedback from those who had attended, and Esvelt himself.

Esvelt's plan was to attack the disease earlier in its food chain than when it latched on to and was carried around by deer. The insect acquired the infectious disease as larvae from feeding on white-footed mice, four-inch rodents that live all over the contiguous United States. Esvelt's plan was to genetically engineer white-footed mice that were either immune to Lyme disease, and thus couldn't spread it to the tick larva, or resistant to the tick's saliva, which allowed the tick to attach to and feed on the infected mice, thus cutting off the disease cycle before it began.

Esvelt believed that around three hundred thousand genetically modified mice released on Nantucket would make a serious dent in the incidents of Lyme disease by overtaking the indigenous rodent population, interbreeding, and eventually becoming the dominant, and perhaps the only, mouse living on the island. Ting admitted, the visual image — more than a quarter million genetically altered mice skidding along the cobblestones on their tiny feet, disappearing into basements, rain gutters, into the underbrush to combat Lyme disease — was a little troubling, but to her surprise, the general mood of the meeting had been open and accepting. Esvelt had been careful to explain that his team planned to test the idea on an uninhabited island first — akin to Church's genetically enhanced mosquitoes being tested under domed villages. Only then, when it was proven they'd fit in with the island's ecosystem, would the mice be released onto Nantucket. But even still, the implementation of the program had ethical implications.

In recent years, Ting had dedicated herself to community outreach such as this. She'd traveled to cities all over the country speaking to people, often in impoverished communities, about genetics and the possibilities of genomic engineering. Going out into the civilian population to connect with people about the changes that would affect every one of us had become as important as any of the scientific work she was pursuing. As Church had said many times, science doesn't take place in a vacuum; scientists need to be open about their work and communicate.

The transgenic mouse that Esvelt was

proposing had implications far beyond the little island of Nantucket. The transgenic mouse was really the first step toward a "gene drive" to rid the entire Northeast, and the rest of the world, of Lyme. Eventually, it wouldn't just be changing a gene in a few hundred thousand white-footed mice to make them immune to the disease; the next step would be to insert the genetic change into the species line, to change the entire species.

Gene drives — a change in a gene that was passed to all of a living thing's heirs — were controversial. Changing genes that were inherited meant changing a species, and gene drives could just as easily be used to *end* a species. Rather than making the white-footed mice unable to carry Lyme disease, Esvelt could have proposed to make the mice unable to reproduce.

In the realm of mosquitoes, that exact solution was being worked on by multiple private companies. In fact, in Brazil one such project was already being tested in the field to combat dengue fever and, eventually, the Zika virus: A quarter million *Aedis aegypti* mosquitoes created by a UK company called Oxitec had been released into a village in São Paulo; these particular mosquitoes had been created using a gene drive that led to highly competitive males who could live only four days, and whose larvae couldn't survive to adulthood. The genetically altered mosquitoes had overwhelmed the local natural population, and cases of dengue had dropped to a tenth of what had previously been reported.

In short, the genetic modification had been successful, and in that local area, the *Aedis aegypti* species was on its way to being gene-driven to extinction. Although most people would argue that the elimination of disease-carrying mosquitoes was for the greater good, there were still questions about the ethics of using such a powerful scientific tool. In fact, U.S. intelligence officials had recently determined — causing much controversy in the science world — that CRISPR and gene drives should be considered potential weapons of mass destruction: Altering genes in a species line could cause almost immediate genocide of entire species.

But the people of Nantucket saw that the science of gene drives, applied in a beneficial way, could rid them of a disease that threatened their health and their bank accounts.

"I'm the first person to say if you go tinker with Mother Nature, we're going to break it," one of the gathered townspeople, as reported by the *New York Times,* had told the floor, encapsulating the mood. "But you know what? Even I want to see where you go with this."

Everything about the presentation — the openness, the involvement of the community, the methodical plan laid out for implementing the idea — was exactly what Ting considered science done right. Science in secret was dangerous, difficult to regulate, and people who relished secrecy usually had something to hide. Her husband had proven that it was possible to be competitive, to break new ground, without locking down your lab behind opaque walls, towers with armed guards, and fences of barbed wire.

And in science such as this, it was especially necessary to involve the greater public. Every person had a stake in something as powerful as gene drives, whether in mosquitoes or in mice.

Though she knew Church felt similarly, she guessed he was much more excited about the results of the town hall meeting than about the process. A fleet of transgenic mice ending Lyme disease was the sort of real-world solution to a health problem with genetic science that his lab aimed to provide. He also felt that the ethics, safety, and communication had to be done very well.

Certainly, he did not fault the community for initially being wary; there were real dangers behind such powerful uses of genetic engineering. The general public had a right to ask questions. Church himself was no stranger to such debate.

During his appearance on the *Colbert Report* in 2012, the host had asked him point blank: "How do you think your work will eventually destroy all mankind? Do you think it's going to be like a killer virus, or more like a giant mutant killer squid man?"

Although Colbert had obviously been joking, the sort of science Church did on a daily basis pushed boundaries, and sometimes when one moved beyond boundaries, one ended up in realms brimming with significant risks.

Only recently, Church and a team in his lab had created a mostly synthetic life-form: They had created the first "recoded" or "genetically synthetic" organism, a strain of E. coli with a radically changed genome. It was the first example of genome-scaled engineering — not just the addition of genes.

The experiment had taken place in a sterile environment with highly redundant safety precautions. From air filtration to sterile, contained work spaces, the environment was completely controlled, and throughout their work Church and his team wore the requisite biosafety gear — gloves, masks, and sealed scrubs. This wasn't a Hot Zone, and the work didn't involve Racal bodysuits with oxygen tanks, but everyone involved understood the inherent danger.

Contamination in any lab could happen quite simply. Lean over to pick up a dropped pipette or Petri dish, tear your gloves or your sleeve on the edge of a cabinet; a microscopic sample touches skin, and a brand-new, synthetic bacteria begins its journey to the outside world. Stop in a coffee shop on the way home, go see a movie, and suddenly a form of E. coli that doesn't exist in nature — because, as Colbert foresaw, it was invented in a lab — has entered the biosystem of the City of Boston. It travels from a movie theater seat or a poorly washed coffee mug to a stranger, who takes it home via the subway.

No immune system in the city, in the state, in the country, in the world has any built-in defense against a synthetic bacteria, because it hasn't existed before. Maybe, most likely, it's harmless; but maybe it's not.

Church understood that no matter how perfectly safe a physical environment felt, no matter how powerful a reverse ventilation system was or how many gloves and bodysuits a scientist wore, there was always the possibility that something could go wrong. A test tube could shatter, a ventilation system could break down. The key to true safety was to build measures into the work itself, a process called "biocontainment." Church had often likened it to putting in seat belts and air bags when one manufactured a car.

In the case of Church's E. coli, he had implanted a synthetic amino acid in the microbe's genome, which it simply could not survive without. Since the amino acid did not exist naturally, and wasn't something the bacteria could ever create on its own, the E. coli could live and propagate only where the amino acid existed — which was within Church's lab. To Church, it was an even more effective method than the more commonly used precaution of engineering a "kill switch" into the microbe — usually some sort of susceptibility to a readily available toxin, so that in the case of an accident, the microbe could be quickly eliminated. Church believed a microbe might find a way around such a kill switch, by evolving to tolerate whatever toxin was chosen, but it was not feasible that a microbe could somehow get around an existential need for a synthetic amino acid built into its genetic structure.

Church and his colleagues had gone to extreme lengths to test and perfect their biocontainment strategy. He and colleague Dan Mandell had gone through huge stacks of Petri square plates in order to get to "zero in a trillion" escapes. Another colleague, Michael Napolitano, had built an automated "morbidostat," with big reservoirs of fresh growth media in which they slowly lowered the amounts of the "addicting" chemical. As time passed, the cells in the growth chamber became more and more capable of growing with lower levels of the amino acid, but were never able to grow without some level of the synthetic chemical, proving they could not live on their own.

Would all scientists conduct their experiments with such an unwavering sense of responsibility? Church couldn't be sure. More and more, Church's sort of science didn't require the state-of-the-art labs found at Harvard. At some point, genetic engineering would be possible in garages and attics — if it wasn't already.

Esvelt was going to make his genetically engineered mice in a safe lab at MIT, but one day, Church could imagine, there would be twentysomethings making transgenic mice in the basements of their suburban homes.

All the more reason, Church believed, that it was important for scientists to reach out to the general public, to explain what they were doing, and that the world needed to take notice, to understand.

Church's thoughts on the return ferry were not only about Esvelt's mice. Among other things, he was picturing the town hall meeting he'd one day pull together to unveil the first baby Woolly Mammoth. He intended to be as open and public with the project as he could. Especially as he moved closer and closer to his goals.

He'd recently received his grant to attempt to build a synthetic womb. And Luhan and Bobby were on the verge of implanting synthetically sequenced Mammoth genes into immortalized elephant cells. They had put together fourteen in all — ten more than the original four they had planned — and would soon be attempting to stimulate the cells to make them iPSCs, so they could create organoids and test their results. If they succeeded, there would be a clear path to that first Woolly Mammoth.

But right now, he was on the ferry, and another animal had come into view.

The seagull had reappeared, once again

suspended in the air in front of them.

"Your friend is back," Church said, touching Ting's shoulder with his big hand.

Perhaps, Ting thought, he saw it now as a gull, not as a sum of its parts.

Church always had his feet in both the present and the future, and only he knew for sure which was reality, and which was still just a dream.

CHAPTER TWENTY-EIGHT

Today

77 AVENUE LOUIS PASTEUR, BOSTON.

Ten minutes past two in the morning.

"It's alive," Luhan said, as she watched Church step back from the Petri dish filled with altered hemoglobin cells. The condensation from the cryogenic device swirled around their corner of the lab, only adding to the atmosphere.

Luhan didn't understand why Bobby, standing next to Church in the dim light from the fluorescent tubes high above the work hoods, seemed to be having difficulty holding back a smile at her words; she wasn't aware that she'd referenced the classic 1930 horror film *Frankenstein,* and she hadn't meant to imply that they had just created life. Because they hadn't. Nor was "alive" really the proper description.

"I mean, they're functioning properly. Releasing oxygen."

If the cells had been traveling through the highways and byways of a mammalian circulatory system, rather than floating in a culture trapped in a plastic Petri dish, they would have been feeding the oxygen they carried to peripheral limbs and organs, to skin wrapped around the little bones that made up a tail, a trunk, ears, and padded toes.

"So now we've got an elephant who can live in the Arctic Circle," Bobby said, finally letting his smile grow.

"We've got a lot more than that," Luhan said.

She watched as Church shifted his attention to the row of other Petri dishes that they'd just removed from the sterile refrigeration unit near their work space. Lined up together along the counter, it was a collection that, in Luhan's opinion, ought to have been in a display case front and center at the Smithsonian, or perhaps more appropriately in the offices of *National Geographic.*

Fourteen organoids (ideally as accurate as possible in three-dimensional shape and molecular details) made up of living, immortalized, and chemically rendered "stem-like" Asian elephant cells originally donated by Ringling Brothers Conservation Center, implanted with synthetically created Woolly Mammoth genes by means of CRISPR. Each of the organoids, representing an isolated Mammoth gene, cells dividable beyond the Hayflick limit and ready for testing for the traits those genes were designed to generate.

Woolly Mammoth hemoglobin, Woolly Mammoth subcutaneous fat, Woolly Mammoth ear cells, Woolly Mammoth tail cells — fourteen of the twentysomething traits they eventually hoped to code for. Church was looking through them one by one, going over the testing procedures they'd developed to check their work; in some ways, though the hemoglobin was one of the more difficult genes to sequence, it had been one of the easier to test. You didn't need to grow an ear to see if you'd gotten hemoglobin right.

Luhan and Bobby had been over their test protocols many times in recent weeks. Working without Quinn and Margo hadn't been easy, and they were hopeful that at least Quinn would return at some point in the near future. Soon, Luhan knew, her transgenic pigs would take her away from the Church Lab as well; the company she and Church had founded was already raising significant money, and would be beginning trials with liver material that could lead to their first actual transplantation.

It was bittersweet to think about. To Luhan, graduating from the Church Lab was like stepping off a cloud and into the real world. She knew the Woolly Mammoth Revival project would be in good hands with Bobby, who was still continuing with his aging work as well, but day to day, she would be running a business as much as she'd be challenging the boundaries of science.

Church somehow managed to keep pushing those boundaries ever farther; the postdocs who would undoubtedly take Luhan's place would find that they were stirring up no end of controversies. The reason Church was seeing the organoids for the first time that evening was that he'd aroused another controversy, tangentially related to the science that would soon give them their first Woolly Mammoth.

Not long ago, the *New York Times* had published an article revealing that a private meeting involving 130 of the world's top biologists, ethicists, and chemists had taken place at Harvard Medical School, in order to study the possibility of creating human cells from scratch. That is, they would create the human cells without a donor, completely synthesized from the ground up.

Initially, the project was called HGP2: The Human Genome Synthesis Project, and was meant to be an extension of the original Human Genome Project and Church's Personal Genome Project, except this time, rather than read the DNA code that was the basis for human life, or publicize individual genomes beginning with George Church's, now the goal was to *write* that sequence, to create it, in a lab, using chemicals. Essentially, HGP2 could, according to some of the press coverage, lay the groundwork for making a person without the need for biological parents.

The article had led to an immediate uproar that dwarfed the furor over Church's comments about a possible Neanderthal revival. Ethicists warned about the dangers of playing God, of designer babies birthed in test tubes, of gene drives and the terrifying ramifications that could arise from genetic tinkering in the wrong hands. No matter that Church and others had already created life in a Petri dish — in 2010, Craig Venter had synthesized a tiny mycoplasma bacteria with a million base pairs of DNA, and then from 2013 to 2016 Church himself had designed and revised his version of the E. coli bacteria, complete with its genetic biocontainment strategy.

If creating life were the barometer, Luhan supposed, Church and others had already been playing God for some time. But even she agreed that human cells were different. A pig with a human-compatible liver raised eyebrows, but a human liver from scratch would be a game changer.

While other outlets took the *Times*'s reporting on a "private" meeting to mean it was in some sense a secret conclave, the Harvard Medical School meeting hadn't been any more secret than most meetings. More than 350 invitations were sent out, with no request for confidentiality, and the whole meeting was videotaped and put online for open access. Church and his colleagues had been on the verge of submitting a paper summarizing their research. Before the publication of any noteworthy scientific study, authors request that the press delay reporting until the paper is publicly available. But the science for creating a synthetic human was worthy of headlines, even though they were discussing cells from many species, not human babies.

Church felt that the press and public misunderstood the purpose of the project: It wasn't actually to build a human, just human cells. The availability of these cells to scientists would make research into human diseases such as cancer and Alzheimer's, as well as conditions such as aging, cheaper and more practical.

The estimated price tag for funding Church's new project was $100 million. To Luhan, it seemed a paltry amount, considering that the original Human Genome Project had cost almost $3 billion to complete. More than anything, the controversy around Church's private meeting stemmed from the fear that maybe science was now moving *too* fast, accelerating beyond people's ability to control it.

But closing your eyes won't stop the inevitable use of technology that is already here. Today — not tomorrow — a living cell could be made, synthetically, in a lab. And that was the real holy grail of biology — not the HGP, the reading of the genetic sequence, but the writing of the genetic sequence.

Church finished looking over the assembled Petri dishes with the hemoglobin, then turned back to Luhan and Bobby. Unlike his synthetic human cell project, this was not creating life from scratch, but to Luhan it was just as important. The sample in front of them was no longer the blood of an elephant. It was the blood of a ten-thousand-year-old Woolly Mammoth, revived.

Then Church glanced back at the row of Petri dishes.

"There's something missing, isn't there?"

Luhan's eyes met Bobby's.

"We were wondering when you'd notice."

"It ended up being much more complicated than we'd initially thought," Bobby said, as they walked Church deeper into the lab, passing dormant workstations and humming ventilation hoods. "But we didn't really need to reinvent the wheel. We started with a pipeline based on research done at the University of California."

"They had it a little easier," Luhan continued, as they passed a pair of postdocs bent over an electrophoresis machine, working on separating DNA from cells. "They were able to take stem cells that they'd edited for the proper lineage. But from there, the process was the same."

"We seeded our iPSCs with the inserted gene into a gel matrix," Bobby said, clearly excited. His glasses kept slipping down his nose as he tried to stay a step ahead of Church. "We had to make the matrix ourselves. It was tough getting the nutritional level just right. Once it had been seeded in, and started to thrive, we peeled the sample out and were ready for the graft procedure."

They had grafted the cells into mice. Luhan didn't like trials involving live animals — whether nude mice or transgenic pigs — but sometimes, it was necessary. Sooner or later, experiments needed to leave the Petri dish.

The three of them headed toward a plastic and glass animal cage sitting on a counter beneath one of the work hoods. Luhan had done the procedure herself on a little nude mouse, a mutant breed, born without an immune system. The little mouse had been lying unconscious on its side, its tiny, hairless chest rising and falling as the anesthetic moved through its bloodstream. She had used a scalpel to make a small incision on the rodent's back, then employed a tiny syringe to implant the slice of gel into the open wound. It had been delicate work, suturing the area closed. The mouse was much smaller and more fragile than the pigs she worked with.

"How long?" Church asked, as they approached the cage. As they moved closer, Luhan could just make out, through the transparent cage walls, the little figure of the mouse, now healed and wide awake, running inside a metal wheel. The mouse's body was mostly a blur as it worked the wheel, which gave off a rhythmic creak, metal against metal, the metronome of a happy rodent in motion.

"Almost two months," Luhan said. "We started to see results at five weeks, but waited for it to become more pronounced."

Church looked at her, then hurried his step, making short work of the distance between the three of them and the mouse on the wheel.

Up close, Luhan noted with satisfaction that the mouse was indeed healthy, tiny paws turning the wheel with a minimum of effort. Her surgical technique had been adequate, and the scar on the mouse had entirely healed.

"It's . . . stunning," Church said, his voice more breath than words.

Luhan wondered if anyone had called a nude mouse stunning before. But the truth was, this nude mouse was no longer completely worthy of its name. Though most of its body — its stomach, head, hindquarters, and tail — was still hairless, one area was decidedly . . . *not.*

The area where Luhan had performed her delicate surgery — the area where she had inserted the engineered cells they had grown in their nutritional matrix — wasn't naked at all.

It was covered in bright red hair.

Chapter Twenty-Nine

Three years from today

TWENTY KILOMETERS SOUTH OF WRANGEL ISLAND.

At ten minutes past three in the afternoon, Nikita Zimov braced himself against the heavy waves as he stared at the canopy of dark clouds, hovering so low they seemed to blend right into the heavy fog. The fog was laced with a cold drizzle, the kind of weather that stung right through clothes and skin, each drop seeming to dig right into bone.

The ocean was rough, spraying against the sides of Nikita's boat. All around, the ocean was pocked with oddly shaped ice floes, some only a few feet in length, others even bigger than Nikita's vessel. But when Nikita leaned forward to look over the railing, the surface of the water was clear, a gorgeous shade of blue, the kind of water he'd expected to see in the Caribbean, around an island brimming with rich businessmen and beautiful women in bikinis.

He laughed at the thought. The kind of people who visited Caribbean islands wouldn't get anywhere near a place like this.

He stepped back from the railing, looking toward the stairs that led down to the small cabin of the thirty-foot cruiser. If it wasn't for the fog, he could have stood on his toes and maybe caught sight of the twenty-foot-long, high-speed Yamaha tethered behind the cruiser — a makeshift tender/lifeboat, because you never really knew what you were going to run into this far north. Scratch that: You had the lifeboat because you knew exactly what you'd run into this far north. Misery, hardship, danger.

Both boats were packed to the gills with supplies. Although the shortest route from Chersky to Wrangel Island was about nine hundred kilometers, most of that had been spent traversing the Kolyma River; once you exited the mouth of the Kolyma, it was another two hundred kilometers through the open Arctic Ocean. A trip for a madman, Anastasiya had called it, something Nikita had repeated numerous times with no small amount of pride. True, it wasn't a trip for the lighthearted, but nobody in Nikita's family had ever been accused of suffering from that particular affliction.

In fact, Nikita had made this trip once before, back in August 2010. On that journey, he had been accompanied by his father, Sergey, who had been fifty-five at the time, his uncle Victor, fifty-eight, and a twenty-year-old family friend named Alexei who had worked for them at Pleistocene Park. Of the four of them, it had been the twenty-year-old who had suffered the most.

On that trip, as well as the current journey, they had gone to retrieve cargo for the park. In 2010, they had made the crossing for musk oxen: Thirty-five years earlier, a small herd of the mixed species — basically, a cross between an ox, a giant sheep, and a goat — had been brought to Wrangel where villagers had raised them for food, milk, and labor. Sergey Zimov had realized that the musk ox — sturdy, horned, covered in dark fur — was perfectly suited for their growing habitat, and had made a deal to retrieve six to ten of the beasts to help populate Pleistocene Park.

Nikita's memories of that earlier trip to Wrangel were mostly fond, though it had been an intense learning experience. It had taken an extra few days at sea to find entrance to the ice-locked island, and by the time they'd worked their way past the ice floes, through bouts of engine trouble and storm damage, they had been thoroughly exhausted. The way back had been even harder, dealing with the elements while tending to a family of musk oxen who didn't particularly like living in the hull of a boat. It hadn't been as bad as carting a group of angry elk across the Russian continent, but surviving the journey had involved copious amounts of vodka that Nikita had kept hidden in a diaper box his wife had left on board from a previous trip downriver with one of their children.

This current trip, too, there had been plenty of need for vodka. Things had started off fine: The first ninety kilometers down the Kolyma had been calm and quiet. Barely any waves, the two boats cutting through the glassy water like a sharp razor through his father's long beard.

When they'd passed the final seaport at the end of the river they had celebrated — again, vodka, but this time shared between Nikita and his partner on the journey, a young Siberian native who worked with Nikita and his father at the Science Center — and by the third day, they'd gone half the distance to the island. They'd even spent time admiring an ice floe, supporting in the center a single happy and fat seal, sunning himself without a care in the world.

It wasn't until they'd reached thirty kilometers outside Wrangel that they'd hit their first major snag; as with Nikita's last trip, they'd found the island surrounded by thick, impassable ice.

That first night, they'd had to anchor the pair of boats and wait for a break in the ice. By the next morning, they'd been able to make a little forward progress, but again had been slowed by ice and forced to tie up for a second night.

The next morning, they'd awakened to a spectacle. Right next to where they'd anchored, a colony of walruses had gathered. The giant animals seemed quite magnificent, though he doubted there was enough vodka in the world for him to mistake them for sirens or mermaids, as ancient sailors supposedly had.

Along with the walruses, the ice kept them trapped again for the next day, which Nikita spent curled up in the main cabin on one of the four small beds. On the previous trip, two of the beds had been covered by baled hay, ready for the musk oxen. On that trip, Nikita and his father had slept in sleeping bags, surrounded by the animals' dinner.

This time, Sergey had remained back in Chersky, in preparation for the return of the cargo. Sergey was half expecting Nikita to return empty-handed. Sergey had seen the same emails as Nikita, but still he didn't quite believe there was going to be anything waiting for them when they reached the frozen island.

Sergey Zimov was a complex man. He was a staunch believer in science, but he lived in the now. He and Nikita had kept current with all the developments they could, from their isolated Arctic laboratory. They knew all about the South Koreans and their Russian partners, working on Mammoth clones. Nikita had seen many frozen carcasses in his journeys around the steppes. The Mammoth graveyard had been a huge economic boon to the area. In fact, at this point, in this part of the world, there were only two types of work for most of the tribes: fishing and recovering Mammoth ivory. Beginning in the early 2000s, ivory hunting had eclipsed fishing, and much of the local population had gone even farther north, hoping to become rich.

Zimov believed that part of the impetus for the shift was cultural: Siberians didn't like to work every day. They preferred to work for a week and then do nothing for the rest of the year. Mammoth ivory provided this option. It had inspired a modern-day gold rush.

Nikita looked at the South Korean effort in similar terms — as taking advantage of a gold rush. Exploiting the Mammoths coming up from the ice, trying to find some way to turn that ancient, dead tissue into gold.

The science that George Church and his lab were conducting was different — genetic engineering, not alchemy. Nikita supported their efforts, hoping for the best.

But the Zimovs weren't in the business of making Mammoths. They weren't trying to revive one species, they were trying to revive an entire ecosystem. The Woolly Mammoth was just one brick in their wall.

After another day of waiting, the walruses left, and, finally, the ice shifted just enough to let them move forward again. After ten kilometers, they'd passed a single ice floe, twice as large as their boat. From the center of the floe, a lone polar bear warily watched them pass. Nikita, once again at the bow of the boat, had instinctively put his hand to his hip, where he kept a large hunting knife. Then he'd grinned at himself: *Crazy Russian.* What would he do with a knife against a polar bear?

Wrangel was teeming with polar bears. On his last trip, his father had nearly shot one of the animals that seemed too close, but

Nikita had stopped him. It didn't seem right to kill a bear when their goal was to repopulate the tundra.

"Better the bear should kill us?" his father had asked with a laugh.

A few hours after they passed the ice floe they found a clear path, ever north.

He started to make out the island, rising up from a new gray fog. On the nearest edge of the shore, a craggy, rocky slope, he could see a heap of rusty barrels. They looked like others he'd noticed on the last trip, which he guessed contained some sort of industrial waste, a reminder of the defunct Soviet Union that had once built bases on remote islands like this. Past the barrels, he saw a pair of tree stumps, traces of storms that had come through, again and again, driving Arctic winds that could reach a hundred miles per hour.

After the stumps, signs of a tiny village of about fourteen houses. One of them contained the barn where, in 2010, he and his father had retrieved the musk oxen. It had been exhausting work to rope the young animals, contain them, and get them aboard the boat.

Today, they hadn't come to Wrangel to rope musk oxen.

"Nikita. Do you see that?"

His Siberian crewmate had come up from the cabin and was pointing past Nikita's shoulder. Nikita followed his finger — beyond the last house of the small village, toward a low slope that ran down to the shoreline. The fog was too thick to see much beyond shapes. Nikita made out what he thought were more tree stumps, and then another handful of barrels. Beyond those, a small wooden construct that might have been a makeshift dock. And then, higher up on the slope, something big and rounded, vaguely reminiscent . . .

Nikita paused, staring.

The shape was moving.

"Nikita . . ."

"I see it," Nikita said, half to himself.

A part of him didn't really want to believe. More accurately, didn't know if he *should* believe.

There was still too much fog to know for sure. Maybe it was something else. A very large musk ox. A walrus or bear.

For the past year, there had been almost no real developments that he and his father had read about, no news sent to their isolated outpost. Most of the news blackout had been their own fault. They did not travel west or east very often, and they didn't communicate with the outside world more than was absolutely necessary. Their English wasn't great, their Korean even worse.

Nikita knew everything he could know about the different approaches, the different teams struggling toward the same future: It was part of being a scientist, the constant battle to be first. First meant accolades, prizes, glory, history, yes, but, more important, first might mean an easier competition for funding for more research. Thus, first could change the world.

In the past, Wrangel had never been a place of firsts. Wrangel was a place of lasts.

And Nikita, like his father, like the island, had never really cared who got there first.

Just that they got there at all.

He held his breath, as the boat moved closer, as the fog began to clear, as the shape grew solid and real . . .

EPILOGUE: BY DR. GEORGE CHURCH

January 24, 2017

ELEVEN KILOMETERS ABOVE EARTH AT −56 DEGREES CELSIUS AND 830 KM/HR.

Head in the clouds, a bit higher than the 8.8 km summit of Mount Everest, flies a "synthetic biologist" augmented by a 737 skin. Inside this biologist resides a 1.4 kg narcoleptic brain dreaming of flying Mammoths (hairy Dumbos) — reflecting on people who ask about flying pigs and the ever-oxymoronic pygmy Mammoths, as if making a routine inquiry about an online consumer product. So what actually stops us from creating these and other fantastical animals? We can start with (but are not limited by) precedents and records so far. The highest-altitude flight for a bird is similar to the 737 heights, with Rüppell's griffon vulture (an endangered species) soaring up to 11.3 km. The fastest animal so far is the peregrine falcon at 320 km/hour, quite fast, but slower than a 737 flight, not to mention that diving is not really flying.

The largest flying animal ever was *Quetzalcoatlus northropi* at 200 kg. It is estimated to have been able to fly for ten days at an altitude of 4.5 km and speed of 129 km/hr. While a conventional pig can get up to 300 kg, the smallest adult mini-pig is 25 kg. So, plenty of room between the 25 kg minimum for pigs and 200 kg maximum for flight. Bat morphogenetic pathways could be moved over to the pig (or elephant) genome to elongate their arms. Another option would be making a bat that looks like a pig — indeed there is a "hog-nosed bat" that would be a good starting point.

Animal swim bladders can generate gas mixtures up to 75 percent oxygen, differing greatly from our normal atmosphere of 21 percent oxygen. Furthermore, many microbes make hydrogen gas as their major metabolic output. So, if an animal grew large, thin bladders filled with hydrogen (rather than air or oxygen), then it could fly like a dirigible. But it might live in mortal fear of thorns and winter static.

So what about flying Mammoths? The Cyprus dwarf elephant (*Elephas cypriotes*) went extinct thirteen thousand years ago. The adults had a weight of only 200 kg. This is at the edge of the *Quetzalcoatlus* limit, and wings could add extra mass. Nevertheless, this is tiny compared to the largest elephant family member (*Palaeoloxodon namadicus*), which weighed in at a whopping twenty-two tons.

Dwarf elephants are an example of a more widespread phenomenon in which many large animals evolve toward miniature versions on islands — while small animals tend to become much larger on islands. Being small is not just about being cute (or flyable), it also tends to track with the rate of development and aging. Mice weigh about one gram at birth and take only twenty days to gestate from fertilized egg to birth. Elephants take twenty-two months to gestate and emerge at 100 kg. Wouldn't it be great for research if the elephant egg-to-birth process took only twenty days?

Many genetics experts (especially those studying natural human DNA variations) will say that body-size traits do not have simple DNA causes or easy manipulation. But the complexity of thousands of natural genetic and environmental factors does not exclude the option of a single manipulated gene having huge impact. For example, increasing one gene for human growth hormone (or the receptor for human growth hormone) results in the largest and smallest human beings and is used clinically to treat a variety of diseases (for example, Turner syndrome, chronic renal failure, Prader–Willi syndrome, intrauterine growth retardation, idiopathic short stature, and AIDS muscle wasting).

An interesting phenomenon kicks in at this point, called Peto's paradox, which is that even though elephants and cetaceans (blue whales at 180 tons) are 100 million times larger than mice and similar creatures (1.8-gram adult Etruscan shrews), the larger animals seem to be much more cancer resistant and resistant to aging. Each time a cell divides (replicates), there is a chance that it will mutate in a way that will make it start dividing without limit, which often is the start of cancer. So 100 million times the number of cells could/should mean that much higher chance of cancer. To some this can be rationalized by the needs of a particular ecosystem niche, such that small, easy prey (like mice) have large litters (twelve) and short gestations (twenty days) so Darwinian selection for long life is scant.

There are many ways to get short life, but what are the mechanisms behind the super-long lives? And can humans benefit from such knowledge? One intriguing new observation from Ting Wu's lab that might intersect with elephant biology is that certain regions of mammalian genomes, appropriately called ultra-conserved elements (aka UCE), are nearly unchanging in normal development and evolution, but highly changeable leading up to cancer. Ting's lab is testing if these UCEs can be harnessed to reduce mutations leading to cancer and aging or radiation health issues in space.

While in the early stages of planning radically new forms of life, we should keep in mind humane treatment of humans and animals. Even though humans and our works are part of nature, "natural" could be defined as prehuman. Natural is not necessarily kind or benignant, as seen herein in chapters 24 and 26 with the elephant endotheliotropic herpes virus (EEHV) or human smallpox, naturally occurring diseases. If we make these viruses extinct, the animals and humans can, arguably, lead happier, longer lives. The same may go for UCEs, pain pathways, and so on. Ideally, all of the intermediate steps between here and there will also be humane. An interesting question is whether such genetically modified organisms will be regulated in the same manner as "transgenics," which are banned in some countries and from "organic foods." The key difference is that transgenics introduce whole genes from distant species, while "cis-genics" are smaller changes that can happen in normal mutations and inter-breeding species. More than thirty such cis-genic GMOs have been approved by the USDA as exempt from the usual transgenic definitions. This may be quite relevant to preventing extinction of elephants and other species and making their lives more pleasant.

Another frequently asked question is: Can we go farther back than the seven-hundred-thousand-year record, so far, for ancient DNA, and in particular, harvest the DNA of dinosaurs? Some information in ancient proteins seems to last much longer than that of ancient DNA. Proteins are studied by electron microscopy, immunological assays, mass spectrometry, and Fourier transform infrared spectra. The oldest such evidence is from 195-million-year-old ribs of a sauropodomorph dinosaur, *Lufengosaurus,* reported in 2017. (For perspective, *Tyrannosaurus rex* lived relatively recently — a mere 67 million years ago.) Another option for getting very ancient information is based on comparing the genomes of living species. The oldest of all is from scientists Betül Kaçar, Lily Tran, Xueliang Ge, Suparna Sanyal, and Eric Gaucher, who in 2016 resurrected a 700-million-year-old version of a protein called EF-Tu, inferring its protein sequence from comparisons of DNA from living species.

Could something that looks and acts like a dinosaur (ideally a herbivore) be a few years away? In comparing the genomes and developmental biology of living species of birds and reptiles, with dinosaurs in mind, a good starting point would be the ostrich. The ability to grow teeth and front clawed hands is not far off in terms of developmental potential, either by small mutations in the bird genes or by replacing genes lost by bird ancestors with the equivalents found in living species, such as alligators. Similarly, we are learning how to make featherless bodies with long tails. The scales on the feet of an ostrich are of similar developmental origins to the feathers. Many mutations interconvert related tissue types — for example, antennae convert into legs in fruit fly mutants (in a gene aptly called antennapedia). Old adult mouse cells morph into young embryolike cells by inducing only four genes. These examples argue that mutations could be found and optimized to turn all ostrich-feather-producing cells into scale-producing cells.

In contrast to dinosaurs, we have excellent information on DNA from Woolly Mammoths (which lived 5 million to five thousand years ago). Like much of modern science, this is a team effort, often including team members who have never communicated directly. We are currently using four *Mammuthus primigenius* genomes, one from a 44,800-year-old in northeastern Siberia, one that died 4,300 years ago at Wrangel Island (both from Eleftheria Palkopoulou, Swapan Mallick, Pontus Skoglund, Jacob Enk, Nadin Rohland, Heng Li, Ayca Omrak, Sergey Vartanyan, Hendrik Poinar, Anders Götherström, David Reich, and Love Dalén), as well as two from twenty thousand and sixty thousand years ago (from Vincent Lynch, Oscar Bedoya-Reina, Aakrosh Ratan, Michael Sulak, Daniela Drautz-Moses, George Perry, Webb Miller, and Stephan Schuster). We compare these with three Asian elephant genomes and put them in the context of the more distant relative (and more carefully annotated) African elephant reference genome. We are looking for genetic variations (unique to Mammoths) present in four out of four Mammoth genomes and zero out of three Asian elephant genomes.

Kevin Campbell, Alan Cooper, and their coworkers characterized the Mammoth blood protein called hemoglobin. Starting with the Elephas maximus HBB/D gene (from Opazo, Sloan, Campbell, Storz, below) more than 2,100 bp long, only three changes are seen and only these were needed to resurrect the properties of cold tolerance in oxygen exchange. These three changes are underlined and bold, A to G, G to T, and G to C. Despite the fact that 80 percent of the region (in lowercase) does not code for the hemoglobin protein, all of the changes are the coding regions (uppercase and in brackets). You've made it all the way to this point in the book and deserve to see some real Mammoth DNA (in contrast to fake dinosaur DNA mentioned herein in chapter 9).

ttctgggcctcagtttcctcatttgtataataacagaattg-
gagagtaaattcttaagaggcttaccaggctgtaattcta-
aaaaaaatgcataaataaacttgccaaggcagatgtttt-
tagcagcaattcctgaaagaaacgggaccaggagat-
aagtagagaaagagtgaaggtctgaaatcaaactaat-
aagacagtcccagactgtcaaggagaggtatggctgt-
catcattcaggcctcaccctgcagaaccacaccctggc-
cttggccaatctgctcacaagagcaaaaagggcagga-
ccagggttgggcatataaggaagagtagtgccagctg-

ctgtttacactcacttctgacacaactgtgttgactagca-
actacccaatcagacacc[ATGGTGAATCTGA-
CTGCTGCTGAGAAGACACAAGTC**A**CC-
AACCTGTGGGGCAAGGTGAATGTGAA-
AGAGCTTGGTGGTGAGGCCCTGAGCA-
G]gtttgtatctaggttgcaaggtagacttaaggagggtt-
gagtggggctgggcatgtggagacagaacagtctccc-
agtttctgacaggcactgacttcctctgcaccstgtggtg-
ctttcaccttcag[GCTGCTGGTGGTCTACCC-
ATGGACCCGGAGGTTCTTTGAACACTT-
TGGGGACCTGTCCACTGCTGACGCTG-
TCCTGCACAACGCTAAAGTGCTGGCC-
CATGGCGAGAAAGTGTTGACCTCCTT-
TGGTGAGGGCCTGAAGCACCTGGACA-
ACCTCAAGGGCACCTTT**G**CCGATCTG-
A G C G A G C T G C A C T G T G A C A
A G C T G C A C G T G G A T C C T **G** A
G A A T T T C A G G]gtgagtctaggag-
acactctatttttcttttcactttgtagtctttcactgtgattatt-
ttgcttatttgaatttcctctgtatctctttttactcgactatgttt-
catcatttagtgtttttcaacttataccattttgtattacttttct-
ttcaatattcttcctttttcctgactcacattcttgctttatatc-
atgctctttatttaatttcctacgtttttgctcttgctctcccttc-
tcctagtttccttccctctgaacagtacccaaattgtgcata-
ccacctctcgtccactatttctgcactggggcaaatcccc-
acccctcctccatatgagggttggaaaggactgaatca-
aagaggagaggatcatggtgctgttctagagtatgtgat-
tcatttcagacttgaaggataacttgaatatataaaatc-
aggagtaaatggagaggaaagtcagtatctgagaatg-
aaagatcagaaggtcatagacgagatggggagcaga-
agttactaagaaactgaccattgtggctataattaatcac-

ttaattagttaattaatatgtttgttatttattcacgtttttcattt-
tggtgggagtaaatttgggctagtgtgtgggcaacataa-
atgggtttcaccccattgtctcagaggccaagctggatt-
gctttgttaaccatgtctgtgtatgtatctacctcttccccat-
ag[CTCCTGGGCAATGTGCTGGTGATT-
GTCCTGGCCCGCCACTTTGGCAAGGA-
ATTCACCCCAGATGTTCAGGCTGCCTA-
TGAGAAGGTTGTGGCAGGTGTGGCGA-
ATGCCCTGGCTCACAAATACCACTGA]g-
atcctggcctgttcctggtatccatcggaagccccatttc-
ccgagatgctatctctgaatttgggaaaataatgccaac-
tctcaagggcatctcttctgcctaataaagtactttcagct-
caactttctgattcatttatttttttctcagtcactcttgtggtg-
ggggaagttcccaaggctctatggacagagagctcttg-
tgccttataggaaaagttcaagggaaattggaaaataa-
agggaaccatacacagatattaatgggaacaattctac-
ttcaaaggcataaagattgggaaggtttggcaaatagg-
atactggtactacagggattccatgggcctcaggcctaa-
gacatagccccagggctaactttcagattcaattcca-
gaaattactcacaaaataatgga

We could edit the elephant genome one DNA base pair change at a time (for example with CRISPR), or we could make all three changes at once by swapping in the whole 2,100+ bp gene — or even a million base pairs at once. At least three groups (led by teams at Harvard, the J. Craig Venter Institute, and New York University) are already engineering genome pieces in this "mega-base" range. Because of the larger extent and easier access to radically novel sequences, this is often called genome "writing" rather than "editing." As few as three thousand of these mega-base segments could cover the whole elephant genome, allowing nearly perfect conversion to a Mammoth genome. If all we want is a cold-resistant elephant, then we might want to make only a few dozen small changes, but as the technology gets cheaper we will probably try many versions, including some that are with obsessive-compulsive levels of molecular realism. If we get that good at genome engineering, what about the unsequenced part of genomes, plus the epigenomes and microbiomes?

Experts in ancient DNA reading have been dismissive of the prospects for sequencing a whole genome from ancient DNA. This is a fairly reasonable position today, since no mammalian genome has been completely sequenced even for living species (even for humans), and ancient DNA is shredded into millions of pieces per cell by radiation, which make it even more daunting. Beth Shapiro said in her 2015 book, *How to Clone a Mammoth,* "Because we cannot know the complete genome sequence of an extinct species, synthesizing a complete genome from scratch would not be an option." Svante Pääbo said in his 2014 *New York Times* op-ed piece, "Neanderthals Are People Too": "Since the DNA preserved in ancient bones has degraded into short pieces, we cannot tell from which copies of these repeated sequences they come and so we cannot reconstruct exactly how they were arranged."

But here is another great opportunity for thinking out of the box. Visualize the machines that cleanly punch a two-dimensional cardboard image into a million jigsaw puzzle pieces, all still in place and readable. The machine then shakes up the pieces into a box for sale, and the puzzle becomes very hard to assemble (and then read). So if you can read the puzzle after cutting but before shaking, then maybe reading is easy. We hope to try this idea on a variety of ancient genomes soon, using an amazing new method developed in Ting Wu's laboratory called "Oligopaints" (and related in situ sequencing methods). We can lock the cut pieces in their original places by using chemical "fixatives" and restraining polymers and then scan the clear three-dimensional cells using fluorescent confocal microscopy.

Okay. But haven't we lost the Mammoth microbiomes, and isn't our ability to understand the biomes of viruses, bacteria, and fungi in our bodies too primitive? It should be noted that we have been engineering these invisible communities since the 1500s, with the dawn of smallpox vaccines in China. Today, the engineering of body ecosystems has become so advanced that companies have been founded around the process, such as Seres, SynLogics, AOBiome, Fitbiomics, and Holobiome. We study and use the microbiomes of modern elephants that eat and play in snow as well as other herbivores.

Finally, isn't the epigenome even more perplexing than the missing genomic DNA and the microbiome? Well, we can read important parts of the epigenome via the methylated cytosine bases that persist in ancient Mammoth DNA in a variety of body tissues. We also can leverage the epigenome of Asian elephants, which were probably genetically interfertile with Mammoths, just as breeds of dogs with wildly different traits (for example, nine-hundred-fold range in body size) have compatible epigenomes. Whether using genome writing or editing, we change only a very small fraction of the elephant genome to make it identical to Mammoth DNA (for example, the three out of 2,100 bases in the example above) — and we give plenty of time for the epigenome to spread onto the new DNA in the cells. Also, much of the epigenome is reset during the moving of the engineered genome into the pluripotent embryo or eggs en route to making a fetal GMO elephant.

AFTERWORD: MAMMOTH PLUS BY STEWART BRAND

Reviving and restoring Woolly Mammoths — and their climate-stabilizing Mammoth steppe — is the most spectacular wildlife project that Ryan Phelan and I have taken on for our California nonprofit called Revive & Restore, and thanks to George Church's marvelous team, it is the furthest along in terms of actually editing genes from an extinct species into the genome of a living relative. What they are doing is brilliant, breakthrough science, and therefore exactly the kind of high-visibility, proof-of-concept example that will show conservation biologists and the general public what a potent new toolkit biotech is bringing to wildlife conservation.

Still, restoring Woolly Mammoths all the way back to life and to the wild will take many decades, probably most of this century. Identifying and editing all the right genes for the first round of Mammothlike embryos will take time. Developing a successful artificial uterus will take time. Rearing by Asian elephant parents will take time. An elephant generation takes fifteen years from a newborn female to its sexual maturity and then another twenty-two months to the first daughter. Acclimatizing to the far north will take time (though Asian elephants already love snow, as can be seen at a zoo in Ontario). The multiple stages of release to the northern wild will take time. (Russia? Canada? Both?) Each step of the way will be thrilling news. Each ponderous step.

Easier de-extinctions are also under way at Revive & Restore. If ambitiously funded, the first proxy passenger pigeons could be alive as early as 2022. Since they are sexually mature in just seven months, there could be enough birds to start flocking into the wild by 2032. Along the way, an extinct grouselike bird of the American East Coast called the heath hen will probably be brought back to life as a pioneer species to develop the capability for other birds. (Heath hens are genetically close enough to domestic chickens to build on the sophisticated primordial-germ-cell technology invented for chickens.)

Other extinct species are leading candidates for revival. Tasmania might be able to welcome back its apex predator, a marsupial wolf known as the Tasmanian tiger — hunted to extinction in the 1930s. New Zealand's famous ostrichlike moas might make a comeback. In Europe, the impressive mother of all cattle, the aurochs, which has been extinct since 1627, could return. The entire northern Atlantic Ocean was once fished by a flightless, penguinlike bird, the great auk, until the last ones were killed by 1852. They might be revived via their close relative, the razorbill. Other candidates in North America are the colorful Carolina parakeet (extinct by 1918) and the presumed-extinct "Lord God" bird, the ivory-billed woodpecker. The DNA for all these species is well preserved in museum specimens.

De-extinction is dramatic, but it represents only a small part of the benefits that genomic technology can bring to wildlife conservation. Many remnant populations of animals in the wild and in captive breeding programs are facing what is called an "extinction vortex," as inbreeding forces them into an accelerating loss-of-fitness spiral. America's most endangered mammal, the black-footed ferret, might be approaching that situation. That's why Revive & Restore is exploring with U.S. Fish &

Wildlife, the San Diego Frozen Zoo, and the biotech company Intrexon the possibility of cloning back to life two ferrets whose tissue was cryopreserved thirty-five years ago. When revived and breeding, they will enrich the ferret gene pool by increasing the number of founders of the current population from seven to nine. Yet further genetic diversity might be mined from museum specimens that harbor healthy gene variants (alleles) now missing from living ferrets. If this approach works, it could be applied to a variety of endangered species that need their gene pools restored to healthy variability.

Many wild populations of animals and plants are profoundly threatened by exotic diseases — chytrid fungus in frogs, sylvatic plague in black-footed ferrets, Rapid Ohi'a Death in the keystone ohi'a trees of Hawaii, avian malaria in the forest birds of Hawaii. Can disease resistance be engineered into the genomes of those species? It has been done successfully for the legendary American chestnut tree, once driven to functional extinction by the blight that killed 4 billion trees in the early twentieth century. Scientists at SUNY in New York made the trees blight-proof by introducing a fungus-resistant gene from wheat, and the improved tree is now going through the approval process with government regulators. A different approach could work for avian malaria in Hawaii. There the exotic disease is carried by an alien invasive vector, the *Culex quinquefasciatus* mosquito, which also transmits a human disease, West Nile virus. Several existing genetic techniques could be used to eliminate the mosquitoes from the islands, thereby protecting all the birds (and humans) at once.

The term that covers all of these projects (and more to come) is "genetic rescue." In normal times, wild populations would evolve around such problems, but humans are introducing so many challenges so rapidly that evolution doesn't have time to generate the needed adaptations. Conservation biologists call what we are doing "facilitated adaptation." It consists of careful genomic analysis, then minimal gene tweaking, followed by sustained monitoring at every level from ecosystem to individual gene. The goal is to restore ecological biodiversity via precisely enhanced genomic biodiversity.

At Revive & Restore, we find that people come for the Mammoths, but they stay for the ferrets and frogs and trees and birds that need help right now. You'll find more information at our website: reviverestore .org. You are welcome to bring your skills or your resources to the projects you find there.

BIBLIOGRAPHY

Abbasi, Jennifer. "Pioneering Geneticist Explains Ambitious Plan to 'Write' the Human Genome." November 2016. *JAMA.*

Abbot, Alison. "The quiet revolutionary: How the co-discovery of CRISPR explosively changed Emmanuelle Charpentier's life." April 2016. *Nature.*

Austen, Ben. "Stewart Brand: The Last Prankster." March 2013. *Men's Journal.*

Baer, Jake. "This Korean Lab has nearly perfected dog cloning, and that's just the start." September 12, 2015. *Business Insider.*

BEC Crew. "150 Scientists just met in secret to discuss creating a synthetic human genome." May 16, 2016. Science alert.com.

Bretkelly, Jody. "Old Bunkhouse now welcomes both human guests, birds." February 5, 2015. Sfgate.com.

Brown, Katrina. "Mammoth Jurassic Park may be under development in Northern Alberta." March 27, 2014. Imgism.com.

Church, George. *Regenesis: How Synthetic Biology Will Reinvent Nature and Ourselves.* April 2014. Hachette Book Group.

Cyranoski, David. "Cloning Comeback." January 14, 2014. *Nature.*

Dean, Josh. "For 100,000, You Can Clone Your Dog." October, 22, 2014. Bloomberg.com.

Dutchen, Stephanie. "No Escape." January 21, 2015. *HMS* (Harvard Medical School) *News.*

"Elephants Learn from Others." Elephant voices.org.

Grant, Bob. "Credit for CRISPR: A Conversation with George Church." December 29, 2015. *The Scientist.*

Hall, Yancey. "Coming Soon: Your personal DNA map." March 7, 2006. *National Geographic.*

Harmon, Amy. "Fighting Lyme Disease in the Genes of Nantucket's Mice." June 7, 2016. *New York Times.*

Hays, Brooks. "Woolly Mammoth DNA successfully spliced into elephant genome." March 25, 2015. UPI.com, *Science News.*

Honeyborne, James. "Elephants Really Do

Grieve Like Us." January 30, 2013. Daily Mail.com.

Kalb, Claudia. "A New Threat in the Lab." June 9, 2005. *Newsweek.*

Kazutoshi Takahashi, Koji Tanabe, Mari Ohnuki, Megumi Narita, Tomoko Ichisaka, Kiichiro Tomoda, Shinya Yamanaka. "Induction of Pluripotent Stem Cells from Adult Human Fibroblasts by Defined Factors." November 30, 2017. http://ac.els-cdn.com/S0092867407014717/1-s2.0-S0092867407014717-main.pdf?_tid=492ac6ac-2e8e-11e7-adc3-00000aa cb360&acdnat=1493657610_316226a04082e8a7db2290a1252e6bf4.

Klinghoffer, David. "An Apology for Harvard's George Church (of Neanderthal baby fame?)." January 23, 2013. EvolutionNews.org.

Larmer, Brook. "Of Mammoths and Men." April 2013. National Geographic.com.

Lewis, Danny. "Last Woolly Mammoths Died Isolated and Alone." May 8, 2015. *Smithsonian Magazine.*

Lewis, Tanya. "Woolly Mammoth DNA Inserted into Elephant Cells." March 26, 2015. Livescience.com.

Miller, Peter. "George Church, the Future Without Limit." June 2014. *National Geographic.*

Mullin, Emily. "Obama advisors urge action against Crispr Bioterror threat." November 17, 2016. *MIT Technology Review.*

Nickerson, Colin. "A quest to create life out of synthetics." April 2, 2008. *Boston Globe.*

Pollack, Andrew. "Jennifer Doudna, a Pioneer Who Helped Simplify Genome Editing." May 11, 2015. *New York Times.*

Pollack, Andrew. "Custom-made Microbes, at Your Service." January 17, 2006. *New York Times.*

Saletan, William. "The Healer." October 2012. Slate.com.

Scudellari, Megan. "How IPS cells changed the world." June 15, 2016. *Nature.*

Seligman, Katherine. "The Social Entrepreneur: Ryan Phelan's controversial new venture . . ." January 8, 2006. Sfgate.com.

Service, Robert F. "Synthetic Microbe Lives with Fewer than 500 Genes." March 24, 2016. *Science,* Sciencemag.org.

Shapiro, Beth. *How to Clone a Mammoth: The Science of De-Extinction.* April 2015. Princeton University Press.

Siberian Times. "South Koreans kick off efforts to clone extinct Siberian cave lions." March 4, 2016.

Singer, Emily. "The Personal Genome Project." January 20, 2006. *MIT Technol-*

ogy Review.
Stanganelli, Joe. "Interference; a CRISPR Patent Dispute Roadmap." January 9, 2017. Bio-itworld.com.
Stein, Rob. "Disgraced Scientist Clones Dogs, and Critics Question His Intent." September 30, 2015. *All Things Considered,* NPR.
Switek, Brian. "How to Resurrect Lost Species." March 11, 2013. *National Geographic.*
Tahir, Tariq. "Preserved Woolly Mammoth with Flowing Blood Found for First Time, Russian Scientists Claim." May 29, 2013. Metro.com.
Wade, Nicholas. "Regenerating a Mammoth for 10 Million." November 19, 2008. *New York Times.*
Wade, Nicholas. "2 New Methods to Sequence DNA Promise Vastly Lower Costs." August 9, 2005. *New York Times.*
Wilmut, Ian. "Produce Woolly Mammoth Stem Cells, Says Creator of Dolly the Sheep." August 1, 2013. *Scientific American.*
Wolf, Adam. "The Big Thaw." September 2008. *Stanford Alumni* magazine.
"South Korean geneticists to try to clone extinct Siberian lions." March 6, 2016.

RT.Com.
"The Alta Summit, December 1984." Human Genome Project Information Archive. Web.orni.gov.

ACKNOWLEDGMENTS

First and foremost, I am indebted to George Church, Chao-Ting Wu, and their daughter, Marie, for generously lending me their time and stories; *Woolly,* for me, was a true labor of love, the sort of story I've been looking for all my life. I have enormous respect for Dr. Church and his family, and everything he is doing to make our world a better place. Also enormous thanks to Stewart Brand and Ryan Phelan, true icons, as well as Sergey and Nikita Zimov, who are out there in the tundra, boots in the snow, fighting the good fight every day, for all of us. I am also indebted to the Revival Team — Luhan Yang, Bobby Dhadwar, Justin Quinn, and Margo Monroe — without whom *Woolly* would not have been possible.

I am also indebted to Oscar Sharp; I am privileged to be connected to such a phenomenon at this early stage of what will no doubt be a spectacular career. Also great thanks to Marty Bowen, John Fischer, Daria Cercek, Jono Chanin, and the teams at Fox and Temple Hill who are going to make one hell of a movie out of this story.

I am immensely grateful to Leslie Meredith and Peter Borland, fantastic editors, Daniella Wexler, David Brown, and all the people at Atria/Simon & Schuster who have helped bring *Woolly* to light. I am also indebted to Eric Simonoff and Matthew Snyder, agents extraordinaire. And, as always, I'm thankful to my parents, my brothers, and their families. And to Tonya, Asher, Arya, and Bugsy — all ready and waiting for *Woolly* to one day soon, once again stroll the Siberian plains.

ABOUT THE AUTHOR

Ben Mezrich graduated magna cum laude from Harvard. He has published nineteen books, including the *New York Times* bestsellers *The Accidental Billionaires,* which was adapted into the Academy Award–winning film *The Social Network,* and *Bringing Down the House,* which has sold more than 1.5 million copies in twelve languages and was the basis for the hit movie *21,* and most recently, *The 37th Parallel.* He has also published *Once Upon a Time in Russia, Ugly Americans, Rigged,* and *Busting Vegas,* as well as *Bringing Down the Mouse,* a book for young readers. He lives in Boston.

The employees of Thorndike Press hope you have enjoyed this Large Print book. All our Thorndike, Wheeler, and Kennebec Large Print titles are designed for easy reading, and all our books are made to last. Other Thorndike Press Large Print books are available at your library, through selected bookstores, or directly from us.

For information about titles, please call:
(800) 223-1244

or visit our website at:
gale.com/thorndike

To share your comments, please write:
Publisher
Thorndike Press
10 Water St., Suite 310
Waterville, ME 04901